ISBN 978-1-330-31911-6
PIBN 10025626

This book is a reproduction of an important historical work. Forgotten Books uses
state-of-the-art technology to digitally reconstruct the work, preserving the original format
whilst repairing imperfections present in the aged copy. In rare cases, an imperfection in
the original, such as a blemish or missing page, may be replicated in our edition. We do,
however, repair the vast majority of imperfections successfully; any imperfections that
remain are intentionally left to preserve the state of such historical works.

1 MONTH OF
FREE
READING

at

www.ForgottenBooks.com

By purchasing this book you are eligible for one month membership to ForgottenBooks.com, giving you unlimited access to our entire collection of over 700,000 titles via our web site and mobile apps.

To claim your free month visit:

www.forgottenbooks.com/free25626

Similar Books Are Available from
www.forgottenbooks.com

THE
STEAM ENGINEER'S
HANDBOOK

A CONVENIENT REFERENCE BOOK
For All Persons Interested In

Steam Boilers, Steam Engines, Steam Turbines, and the Auxiliary Appliances and Machinery of Power Plants

BY

International Correspondence Schools

SCRANTON, PA.

1st Edition, 18th Thousand, 3d Impression

SCRANTON, PA.
INTERNATIONAL TEXTBOOK COMPANY

I

PRESS OF
INTERNATIONAL TEXTBOOK COMPANY
SCRANTON, PA.

27463

PREFACE

This handbook is intended as a reference volume for persons engaged in the actual management and care of steam engines, steam boilers, and the auxiliary appliances to be found in the average steam-power plant. The aim of the publishers has been to select data of general interest from the vast store of available material, and to combine therewith information relating to the problems and difficulties likely to be encountered in the daily work of the engineer and the fireman.

In order to keep the book of such size as to be carried conveniently in the pocket, the treatment of many of the subjects has necessarily been brief; but subjects of greater importance, as, for example, the care and operation of boilers and engines, have been dealt with more fully. The various tables have been selected with great care, and only those have been included which are likely to be consulted most frequently. The numerous rules and formulas are stated as simply and concisely as possible, and their applications are clearly illustrated by the full solution of many examples.

Another important feature is the inclusion of abstracts from the license laws of the various cities and states in which such laws have been adopted. These abstracts serve to indicate how and to whom applications for license must be made and what qualifications are necessary in order to obtain licenses of the various grades. The United States marine license law is given in full.

Care has been taken to arrange the several portions in a convenient and logical manner, and a very full index increases further the facility with which any given subject may be located.

This handbook was prepared by Messrs. R. T. Strohm and C. J. Mason, under the supervision of Mr. J. A. Grening, Principal of the School of Steam and Marine Engineering.

INTERNATIONAL CORRESPONDENCE SCHOOLS

Nov. 1, 1912

INDEX

2

The Steam Engineer's Handbook

MATHEMATICS

SIGNS USED IN CALCULATIONS

+ *Plus*, indicates *addition;* thus, 10+5 is 15.

Minus, indicates *subtraction;* thus, 10−5 is 5.

× *Multiplied by*, indicates *multiplication;* thus, 10×5 is 50.

÷ *Divided by*, indicates *division;* thus, 10÷5 is 2.

= *Equal to*, indicates *equality;* thus, 12 in.=1 ft.

Parentheses, ( ), *brackets,* [ ], and *braces,* $\{$ $\}$, have the same meanings, and signify that the operation indicated within them is to be performed first; or, if more than one pair is used, that indicated within the inner one is to be effected first. Thus, in the expression $5(7-2)$, the subtraction is to be made first and the difference then multiplied by 5. Again, in the expression $\frac{1}{2}[7-(3+\frac{3}{4})]$, the addition indicated within the parentheses is to be performed first, the sum thus found is to be subtracted from 7, and then half of the remainder is to be found.

The *vinculum,* ‾‾‾, is used for the same purpose as parentheses, brackets, and braces, but chiefly in connection with the *radical* sign $\sqrt{}$ thus, $\sqrt{}$

The *decimal point* (.) is placed in a number containing decimals, to fix the value of that number; thus, 12.5 is 12 and $\frac{5}{10}$; 1.25 is 1 and $\frac{25}{100}$; and so on.

An *exponent* is a figure written above and to the right of a number to indicate the power to which the number is to be raised.   Thus, 8^2 means that 8 is to be *squared;* that is, 8×8 = 64.   Again, 8^3 means that 8 is to be *cubed;* that is $8 \times 8 \times 8$ = 512.

The *radical sign*, $\sqrt{}$, means that some root of the expression under the vinculum is to be found.   If it is used without a small index figure, it indicates *square root;* thus, $\sqrt{64} = 8$.   The sign $\sqrt[3]{}$ indicates the *cube* root; thus, $\sqrt[3]{27} = 3$.

The signs :   indicate proportion; thus, $3:4::6:8$ is read *3 is to 4 as 6 is to 8.*   Instead of the sign : : the equality sign = is often used; thus, $3:4=6:8$.

The signs ° ′ ″ mean degrees, minutes, and seconds, respectively; thus, 60° 15′ 15″ is read *60 degrees 15 minutes 15 seconds.*

The signs ′ ″ also mean feet and inches, respectively; thus, 7′ 6″ is read *7 feet 6 inches.*

The symbol π (pronounced *pi*) means the ratio of the circumference of a circle to its diameter and has a value, near enough for most practical purposes, of 3.1416.

FRACTIONS

COMMON FRACTIONS

The *numerator* of a fraction is the number above the bar, and the *denominator* is the number beneath it; thus, in the fraction ¾, 3 is the numerator and 4 is the denominator.   Two or more fractions having the same denominator are said to have a *common denominator.*   By *reducing fractions to a common denominator* is' meant finding such a denominator as will contain each of the given denominators without a remainder, and multiplying each numerator by the number of times its denominator is contained in the common denominator.   Thus, the fractions ¼, ½, and ₁₆ have, as a common denominator, 16; then ¼=₁₆; ½=₁₆; ₁₆=₁₆.

By *reducing a fraction to its lowest terms* is meant dividing both numerator and denominator by the greatest number that each will contain without a remainder; for example, in ₁₆, the greatest number that will thus divide 14 and 16 is 2; so that, $\dfrac{14 \div 2}{16 \div 2}$ = ⅞, which is ₁₆ reduced to its lowest terms.

A *proper fraction* is one in which the numerator is less than the denominator, as $\frac{2}{3}$.

A *mixed number* is one consisting of a whole number and a fraction, as $7\frac{3}{8}$.

An *improper fraction* is one in which the numerator is equal to, or greater than, the denominator, as $\frac{17}{8}$. This is reduced to a mixed number by dividing 17 by 8, giving $2\frac{1}{8}$.

A mixed number is reduced to a fraction by multiplying the whole number by the denominator, adding the numerator and placing the sum over the denominator; thus, $1\frac{7}{8}$ becomes, by reduction, $\dfrac{(1\times8)+7}{8}=\dfrac{15}{8}$.

Addition of Fractions or Mixed Numbers.—If fractions only, reduce them to a common denominator, add partial results, and reduce sum to a whole or mixed number. If mixed numbers are to be added, add the sum of the fractions to that of the whole numbers; thus, $1\frac{3}{8}+2\frac{1}{4}=(1+2)+(\frac{3}{8}+\frac{2}{8})=4\frac{1}{8}$.

Subtraction of Two Fractions or Mixed Numbers.—If they are fractions only, reduce them to a common denominator, take less from greater, and reduce result; as, $\frac{7}{8}$ in. $-\frac{9}{16}$ in. $=\dfrac{14-9}{16}=\frac{5}{16}$ in. If they are mixed numbers, subtract fractions and whole numbers separately, placing remainders beside one another; thus, $3\frac{7}{8}-2\frac{1}{4}=(3-2)+(\frac{7}{8}-\frac{2}{8})=1\frac{5}{8}$. With fractions like the following, proceed as indicated: $3\frac{7}{16}-1\frac{13}{16}=(2+\frac{16}{16}+\frac{7}{16})-1\frac{13}{16}=2\frac{23}{16}-1\frac{13}{16}=1\frac{10}{16}=1\frac{5}{8}$; $7-4\frac{3}{4}=(6+\frac{4}{4})-4\frac{3}{4}=2\frac{1}{4}$.

Multiplication of Fractions.—Multiply the numerators together, and likewise the denominators, and divide the former product by the latter; thus, $\dfrac{1}{2}\times\dfrac{3}{4}\times\dfrac{5}{8}=\dfrac{1\times3\times5}{2\times4\times8}=\dfrac{15}{64}$. If mixed numbers are to be multiplied, reduce them to fractions and proceed as shown above; thus, $1\frac{1}{2}\times3\frac{1}{4}=\frac{3}{2}\times\frac{13}{4}=\frac{39}{8}=4\frac{7}{8}$.

Division of Fractions.—Invert the divisor, that is, exchange places of numerator and denominator, and multiply the dividend by it, reducing the result to lowest terms or to a mixed number, as may be found necessary; thus, $\frac{7}{8}\div\frac{3}{4}=\frac{7}{8}\times\frac{4}{3}=\frac{28}{24}=\frac{7}{6}$

= $1\frac{1}{8}$. If there are mixed numbers, reduce them to fractions, and then divide as just shown; thus, $1\frac{5}{8} \div 3\frac{1}{4} = \frac{13}{8} \div \frac{13}{4}$, or $\frac{13}{8} \times \frac{4}{13}$ = $\frac{52}{104} = \frac{1}{2}$.

DECIMAL FRACTIONS

In decimals, whole numbers are divided into tenths, hundredths, etc.; thus, $\frac{1}{10}$ is written .1; .08 is read $\frac{8}{100}$, the value of the number being indicated by the position of the decimal point; that is, one figure after the decimal point is read as so many tenths; two figures as so many hundredths; etc. Moving the decimal point to the right multiplies the number by 10 for every place the point is moved; moving it to the left divides the number by 10 for every place the point is moved. Thus, in 125.78 (read 125 and $\frac{78}{100}$), if the decimal point is moved one place to the right, the result is 1,257.8, which is 10 times the first number; or, if the point is moved to the left one place, the result is 12.578, which is $\frac{1}{10}$ the first number, moving the point being equivalent to dividing 125.78 by 10.

Annexing a cipher to the right of a decimal does not change its value; but each cipher inserted between the decimal point and the decimal divides the decimal by 10; thus, in 125.078, the decimal part is $\frac{1}{10}$ of .78.

Addition of Decimals.—Place the numbers so that the decimal points are in a vertical line, and add in the ordinary way, placing the decimal point of the sum under the other points.

$$\begin{array}{r} 101.257 \\ 12.965 \\ 43.005 \\ 920.600 \\ \hline 1,077.827 \end{array}$$

Subtraction of Decimals.—Place the number to be subtracted with its decimal point under that of the other number, and subtract in the ordinary way.

$$\begin{array}{r} 917.678 \\ 482.710 \\ \hline 434.968 \end{array}$$

Multiplication of Decimals.—Multiply in the ordinary way, and point off from the right of the result as many figures as there are figures to the right of the decimal points in both numbers multiplied; thus, in the example here given, there are three figures to the right of the points and that many are pointed off in the result. If either number contains no decimal, point off as many places as are in the number that does.

$$\begin{array}{r} 21.72 \\ 34.1 \\ \hline 2172 \\ 8688 \\ 6516 \\ \hline 740.652 \end{array}$$

If a product has not as many figures as the sum of the decimal places in the numbers multiplied, prefix enough ciphers before the figures to make up the required number of places and place the decimal point before the ciphers. Thus, in .002 ×.002, the product of $2 \times 2 = 4$; but there are three places in each number; hence, the product must have *six* places, and five ciphers must be prefixed to the 4, which gives .000004; that is, $.002 \times .002 = .000004$.

Division of Decimals.—Divide in the usual way. If the dividend has *more* decimal places than the divisor, point off, from the right of the quotient, the number of places in *excess*. If it has *fewer* places than the divisor, annex as many ciphers to the decimal as are necessary to give the dividend as many places as there are in the divisor; if the dividend is a whole number, annex as many ciphers as there are decimal places in the divisor; the quotient in either case will be a whole number. For example, $\dfrac{25.75}{2.5} = 10.3$; $\dfrac{82.5}{2.75} = \dfrac{82.50}{2.75} = 30$; $\dfrac{7.5}{2.5} = 3$.

Carrying a Division to Any Number of Decimal Places. Annex ciphers to the dividend and divide, until the desired number of figures in the quotient is reached, which are pointed off as above shown. Thus, $36.5 \div 18.1$ to three decimal places $= \dfrac{36.5000}{18.1} = 2.016 +$. (The sign $+$ thus placed after a number indicates that the exact result would be more than the one given if the division were carried further; thus, if the division were carried to six figures, the quotient would be 2.01657.)

Reducing a Decimal to a Common Fraction.—Place the decimal as the numerator; and for the denominator put 1 with as many ciphers as there are figures to the right of the decimal point; thus, .375 has three figures to the right of the point; hence, $.375 = \frac{375}{1000} = \frac{3}{8}$.

Reducing a Common Fraction to a Decimal.—Divide the numerator by the denominator, and point off as many places as there have been ciphers annexed; thus, $\frac{3}{16} = 3.0000 \div 16 = .1875$.

INVOLUTION AND EVOLUTION

SIGNIFICANT FIGURES

In any number, the figures beginning with the first digit* at the left and ending with the last digit at the right, are called the *significant figures* of the number. Thus, the number 405,800 has the four significant figures 4, 0, 5, 8 and the *significant part* of the number is 4058. The number .000090067 has five significant figures, 9, 0, 0, 6, and 7, and the significant part is 90067.

All numbers that differ only in the position of the decimal point have the same significant figures and the same significant part. For example, .002103, 21.03, 21,030, and 210,300 have the same significant figures 2, 1, 0, and 3, and the same significant part 2103.

The *integral part* of a number is the part to the left of the decimal point.

SQUARE AND CUBE ROOTS

By means of the accompanying table, the square, cube, square root, cube root, and reciprocal of any number may be obtained correct always to five significant figures, and in the majority of cases correct to six significant figures.

If the number whose root is to be found contains fewer than four significant figures, the required root can be found in the table, the square root under $\sqrt{n}$, or $\sqrt{10n}$, and the cube root under $\sqrt[3]{n}$, $\sqrt[3]{10n}$, or $\sqrt[3]{100n}$, according to the number of significant figures in the integral part of the number. Thus, $\sqrt{3.14} = 1.772$; $\sqrt{31.4} = \sqrt{10 \times 3.14} = 5.60357$; $\sqrt[3]{3.14} = 1.46434$; $\sqrt[3]{31.4} = \sqrt[3]{10 \times 3.14} = 3.15484$; $\sqrt[3]{314} = \sqrt[3]{100 \times 3.14} = 6.79688$.

In order to locate the decimal point, the given number must be pointed off into periods of two figures each for square root and three figures each for cube root, beginning always at the decimal point. Thus, for square root: 12703, 1'27'03; 12.703, 12.70'30; 220000, 22'00'00; .000442, .00'04'42; and for cube root: 3141.6, 3'141.6; 67296428, 67'296'428; .000000217, .000'000'021'700; etc.

---

*If ciphers are used simply to locate the decimal point, they must not be counted as digits.

There are as many figures preceding the decimal point in the root as there are periods preceding the decimal point in the given number; if the number is entirely decimal, the root is entirely decimal, and there are as many ciphers following the decimal point in the root as there are cipher periods following the decimal point in the given number.

Applying this rule, $\sqrt{220000} = 469.04$, $\sqrt{.000442} = .021024$, $\sqrt[3]{518000} = 80.3113$, and $\sqrt[3]{.000073} = .0418$.

If the number has more than three significant figures, point off the number into periods, place a decimal point between the first and second periods of the significant part of the number, and proceed as in the following examples:

EXAMPLE 1.—Find the results of the following:

(a) $\sqrt{3.1416} = ?$ (b) $\sqrt{2342.9} = ?$

SOLUTION.—(a) In this case, the decimal point need not be moved. In the table under n^2 find $3.1329 = 1.77^2$ and $3.1684 = 1.78^2$, one of these numbers being a little less and the other a little greater than the given number, 3.1416. The first three figures of the required root are 177. $31,684 - 31,329 = 355$ is the first difference; $31,416$ (the number itself) $- 31,329 = 87$ is the second difference. $87 \div 355 = .245$, or .25, which gives the fourth and fifth figures of the root. Hence, $\sqrt{3.1416} = 1.7725$.

(b) Pointing off and placing the decimal point between the first and second periods, the number appears 23.4290. Under n^2 find $23.4256 = 4.84^2$ and $23.5225 = 4.85^2$. The first three figures of the root are 484. The first difference is $235,225 - 234,256 = 969$; the second difference is $234,290 - 234,256 = 34$; $34 \div 969 = .035$, or .04, which gives the fourth and fifth figures of the root. Since the integral part of the number 23'42.9 contains two periods, the integral part of the root contains two figures, or $\sqrt{2342.9} = 48.404$.

EXAMPLE 2.—Find the results of the following:

(a) $\sqrt[3]{.0000062417} = ?$ (b) $\sqrt[3]{50932676} = ?$

SOLUTION.—(a) Pointed off, the number appears .000'006'-241'700, and with the decimal point placed between the first and second periods of the significant parts, gives 6.2417. Under n^3 find $6.22950 = 1.84^3$ and $6.33163 = 1.85^3$. The first three figures of the root are 1.84. The first difference is 10,213. and the second difference is 1,220; $1,220 \div 10,213 = .119$, or .12, which gives the fourth and fifth figures. There is one cipher period after the decimal point in the number; hence, $\sqrt[3]{.0000062417} = .018412$.

(*b*) Replace all after the sixth figure with ciphers, making the sixth figure 1 greater when the seventh figure is 5 or greater; that is, $\sqrt[3]{50932700}$ and $\sqrt[3]{50932676}$ will be the same. Placing the decimal point between the first and second periods gives 50.9327. Under n^3 find $50.6530 = 3.70^3$ and $51.0648 = 3.71^3$. The first three figures of the root are 370. The second difference $2,797 \div$ the first difference $4.118 = .679$ or .68. Hence, $\sqrt[3]{50932676} = 370.68$.

SQUARES

If the given number contains fewer than four significant figures, the significant figures of the square or cube can be found under n^2 or n^3 opposite the given number under n. The decimal point can be located by the fact that if the column headed $\sqrt{10n}$ is used, the square will contain twice as many figures as the number to be squared, while if the column headed $\sqrt{n}$ is used, the square will contain twice as many figures as the number to be squared, less 1. If the number contains an integral part, the principle is applied to the integral part only; if the number is wholly decimal, the square will have twice as many ciphers, or twice as many plus 1, following the decimal point as in the number itself, depending on whether $\sqrt{10n}$ or $\sqrt{n}$ column is used.

To square a number containing more than three significant figures, place the decimal point between the first and second significant figures and find in the column headed $\sqrt{n}$ or $\sqrt{10n}$ two consecutive numbers, one a little greater and the other a little less than the given number. The remainder of the work is exactly as described for extracting roots. The square will contain twice as many figures as the number itself, or twice as many less 1, according to whether the column headed $\sqrt{10n}$ or $\sqrt{n}$ is used. The number of ciphers following the decimal point in the square of a number wholly decimal is determined in the same way.

EXAMPLE.—Find the results of the following:

(*a*) $273.42^2 = ?$ (*b*) $.052436^2 = ?$

SOLUTION.—(*a*) Placing the decimal point between the first and second significant figures, the number is 2.7342, which occurs between $2.73313 = \sqrt{7.47}$ and $2.73496 = \sqrt{7.48}$, found under $\sqrt{n}$. The first three figures of the square are 747. The second difference $107 \div$ the first difference $183 = .584$, or .58. Hence, $273.42^2 = 74,758$.

(*b*) With the position of the decimal point changed, the number is 5.2436, which is between $5.23450 = \sqrt{2.74}$ and $5.24404 = \sqrt{2.75}$, both under $\sqrt{10n}$. The first three significant figures of the root are 2.74, and the second difference $910 \div$ the first difference $954 = .953$, or .95, the next two figures. The number has one cipher following the decimal point, and the column headed $\sqrt{10n}$ is used; hence, $.052436^2 = .0027495$.

CUBES

To cube a number, proceed in the same way, but use a column headed $\sqrt[3]{n}$, $\sqrt[3]{10n}$, or $\sqrt[3]{100n}$. If the number contains an integral part, the number of figures in the integral part of the cube will be three times as many as in the given number if the column headed $\sqrt[3]{100n}$ is used; it will be three times as many less 1 if the column headed $\sqrt[3]{10n}$ is used; and it will be three times as many less 2 if the column headed $\sqrt[3]{n}$ is used. If the number is wholly decimal, the number of ciphers following the decimal point in the cube will be three times, three times plus 1, or three times plus 2, as many as in the given number, depending on whether the $\sqrt[3]{100n}$, $\sqrt[3]{10n}$, or $\sqrt[3]{n}$ column is used.

EXAMPLE.—Find the results of the following:
(*a*) $129.684^3 = ?$ (*b*) $7.6442^3 = ?$ (*c*) $.032425^3 = ?$

SOLUTION.—(*a*) With the position of the decimal point changed, the number 1.29684 is between $1.29664 = \sqrt[3]{2.18}$ and $1.29862 = \sqrt[3]{2.19}$, found under $\sqrt[3]{n}$. The second difference $20 \div$ the first difference $198 = .101+$, or .10. Hence, the first five significant figures are 21810; the number of figures in the integral part of the cube is $3 \times 3 - 2 = 7$; and $129.684^3 = 2,181,000$, correct to five significant figures.

(*b*) 7.64420 occurs between $7.64032 = \sqrt[3]{446}$ and $7.64603 = \sqrt[3]{447}$. The first difference is 571; the second difference is 388; and $388 \div 571 = .679 +$, or .68. Hence, the first five significant figures are 44668; the number of ciphers following the decimal point is $3 \times 0 = 0$; and $7.6442^3 = 446.68$, correct to five significant figures,

(c) 3.2425 falls between $3.24278 = \sqrt[3]{10 \times 3.41}$ and $3.23961 = \sqrt[3]{10 \times 3.40}$. The first difference is 317; the second difference is 289; $289 \div 317 = .911+$, or .91. Hence, the first five significant figures are 34091; the number of ciphers following the decimal point is $3 \times 1 + 1 = 4$; and $.032425^3 = .000034091$, correct to five significant figures.

RECIPROCALS

The *reciprocal* of any number is equal to 1 divided by that number; thus, the reciprocal of 6 is ⅙, because $1 \div 6 = ⅙$. The product of a number and its reciprocal is always 1; thus, ⅛ is the reciprocal of 8, and $8 \times ⅛ = 1$.

The last column of the following table gives the reciprocals of all numbers expressed by three significant figures correct to six significant figures. The number of ciphers following the decimal point in the reciprocal of a number is 1 less than the number of figures in the integral part of the number; and if the number is entirely decimal, the number of figures in the integral part of the reciprocal is 1 greater than the number of ciphers following the decimal point in the number.

EXAMPLE.—Find the reciprocal of the following:

$$(a)\ 379.426; \quad (b)\ .0004692$$

SOLUTION.—(a) .379426 falls between $.378788 = \dfrac{1}{2.64}$ and $380228 = \dfrac{1}{2.63}$. The first difference is $380{,}228 - 378{,}788 = 1{,}440$; the second difference is $380{,}228 - 379{,}426 = 802$; $802 \div 1{,}440 = .557$, or .56. Hence, the first five significant figures are 26356, and the reciprocal of 379.426 is .0026356 to five significant figures.

(b) .469200 falls between $.469484 = \dfrac{1}{2.13}$ and $.467290 = \dfrac{1}{2.14}$. The first difference is 2,194; the second difference is 284; $284 \div 2{,}194 = .129+$, or .13. Hence, $\dfrac{1}{.0004692} = 2{,}131.3$, correct to five significant figures.

n	n^2	n^3	$\sqrt{n}$	$\sqrt{10\,n}$	$\sqrt[3]{n}$	$\sqrt[3]{10\,n}$	$\sqrt[3]{100\,n}$	$\dfrac{1}{n}$
1.01	1.0201	1.03030	1.00499	3.17805	1.00332	2.16159	4.65701	.990099
1.02	1.0404	1.06121	1.00995	3.19374	1.00662	2.16870	4.67233	.980392
1.03	1.0609	1.09273	1.01489	3.20986	1.00990	2.17577	4 68755	.970874
1.04	1.0816	1.12486	1.01980	3.22490	1.01316	2.18278	4.70267	.961539
1.05	1.1025	1.15763	1.02470	3.24037	1.01640	2.18976	4.71769	.952381
1.06	1.1236	1.19102	1.02956	3.25576	1.01961	2.19669	4.73262	.943396
1.07	1 1449	1.22504	1.03441	3.27109	1.02281	2.20358	4.74746	.934579
1.08	1.1664	1.25971	1.03923	3.28634	1.02599	2.21042	4.76220	.925926
1.09	1.1881	1.29503	1.04403	3.30151	1.02914	2.21722	4.77686	.917431
1.10	1.2100	1.33100	1.04881	3.31662	1.03228	2.22398	4.79142	.909091
1.11	1.2321	1 36763	1.05357	3.33167	1.03540	2.23070	4.80590	.900901
1.12	1.2544	1.40493	1.05830	3 34664	1.03850	2.23738	4.82028	.892857
1.13	1.2769	1.44290	1.06301	3.36155	1.04158	2.24402	4.83459	.884956
1.14	1.2996	1.48154	1.06771	3.37639	1.04464	2.25062	4.84881	.877193
1.15	1.3225	1.52088	1.07238	3.39116	1.04769	2.25718	4.86294	.869565
1.16	1.3456	1.56090	1.07703	3.40588	1.05072	2.26370	4.87700	.862069
1.17	1.3689	1.60161	1.08167	3.42053	1.05373	2.27019	4.89097	.854701
1.18	1.3924	1.64303	1.08628	3.43511	1.05672	2.27664	4.90487	.847458
1.19	1.4161	1.68516	1.09087	3.44964	1.05970	2.28305	4.91868	.840336
1.20	1.4400	1.72800	1.09545	3.46410	1.06266	2.28943	4.93242	.833333
1.21	1.4641	1.77156	1.10000	3.47851	1.06560	2.29577	4.94609	.826446
1.22	1.4884	1.81585	1.10454	3.49285	1.06853	2.30208	4.95968	.819672
1.23	1.5129	1.86087	1.10905	3.50714	1.07144	2.30835	4.97319	.813008
1.24	1.5376	1.90662	1.11355	3.52136	1.07434	2.31459	4.98663	.806452
1.25	1.5625	1.95313	1.11803	3.53553	1.07722	2.32080	5.00000	.800000
1.26	1.5876	2.00038	1.12250	3.54965	1.08008	2.32697	5.01330	.793651
1.27	1.6129	2.04838	1.12694	3.56371	1.08293	2.33310	5.02653	.787402
1.28	1.6384	2.09715	1.13137	3.57771	1.08577	2.33921	5.03968	.781250
1.29	1.6641	2.14669	1.13578	3.59166	1.08859	2.34529	5.05277	.775194
1.30	1.6900	2.19700	1.14018	3.60555	1.09139	2.35134	5.06580	.769231
1.31	1.7161	2.24809	1.14455	3.61939	1.09418	2.35735	5.07875	.763359
1.32	1.7424	2.29997	1.14891	3.63318	1.09696	2.36333	5.09164	.757576
1.33	1.7689	2.35264	1.15326	3.64692	1.09972	2.36928	5 10447	.751880
1.34	1.7956	2.40610	1.15758	3.66060	1.10247	2.37521	5.11723	.746269
1.35	1.8225	2.46038	1.16190	3.67423	1.10521	2.38110	5.12993	.740741
1.36	1.8496	2.51546	1.16619	3.68782	1.10793	2.38696	5.14256	.735294
1.37	1.8769	2.57135	1.17047	3.70135	1.11064	2.39280	5.15514	.729927
1 38	1.9044	2.62807	1.17473	3.71484	1.11334	2.39861	5.16765	.724638
1 9	1.9321	2.68562	1.17898	3.72827	1.11602	2.40439	5.18010	.719425
1.40	1.9600	2.74400	1.18322	3.74166	1.11869	2.41014	5.19249	.714286
1.41	1.9881	2.80322	1.18743	3.75500	1.12135	2.41587	5.20483	.709220
1.42	2.0164	2.86329	1.19164	3.76829	1.12399	2.42156	5.21710	.704225
1.43	2.0449	2.92421	1.19583	3.78153	1.12662	2.42724	5.22932	.699301
1.44	2.0736	2.98598	1.20000	3 79473	1.12924	2.43288	5.24148	.694444
1.45	2.1025	3.04863	1.20416	3.80789	1.13185	2.43850	5.25359	.689655
1.46	2.1316	3.11214	1.20830	3.82099	1.13445	2.44409	5.26564	.684932
1.47	2.1609	3.17652	1 21244	3.83406	1.13703	2.44966	5.27763	.680272
1.48	2.1904	3.24179	1.21655	3.84708	1.13960	2.45520	5.28957	.675676
1.49	2.2201	3.30795	1.22066	3.86005	1.14216	2.46072	5.30146	.671141
1.50	2.2500	3.37500	1.22474	3.87298	1.14471	2.46621	5.31329	.666667

n	n^2	n^3	$\sqrt{n}$	$\sqrt{10\,n}$	$\sqrt[3]{n}$	$\sqrt[3]{10\,n}$	$\sqrt[3]{100\,n}$	$\dfrac{1}{n}$
1.51	2.2801	3.44295	1.22882	3.88587	1.14725	2.47168	5.32507	.662252
1.52	2.3104	3.51181	1.23288	3.89872	1.14978	2.47713	5.33680	.657895
1.53	2.3409	3.58158	1.23693	3.91152	1.15230	2.48255	5.34848	.653595
1.54	2.3716	3.65226	1.24097	3.92428	1.15480	2.48794	5.36011	.649351
1.55	2.4025	3.72388	1.24499	3.93700	1.15729	2.49332	5.37169	.645161
1.	2.4336	3.79642	1.24900	3.94968	1.15978	2.49866	5.38321	.641026
1.	2.4649	3.86989	1.25300	3.96232	1.16225	2.50399	5.39469	.636943
1.	2.4964	3.94431	1.25698	3.97492	1.16471	2.50930	5.40612	.632911
1.56	2.5281	4.01968	1.26095	3.98748	1.16717	2.51458	5.41750	.628931
1.60	2.5600	4.09600	1.26491	4.00000	1.16961	2.51984	5.42884	.625000
1.	2.5921	4.17328	1.26886	4.01248	1.17204	2.52508	5.44012	.621118
1.	2.6244	4.25153	1.27279	4.02492	1.17446	2.53030	5.45136	.617284
1.	2.6569	4.33075	1.27671	4.03733	1.17687	2.53549	5.46256	.613497
1.61	2.6896	4.41094	1.28062	4.04969	1.17927	2.54067	5.47370	.609756
1.65	2.7225	4.49213	1.28452	4.06202	1.18167	2.54582	5.48481	.606061
1.	2.7556	4.57430	1.28841	4.07431	1.18405	2.55095	5.49586	.602410
1.	2.7889	4.65746	1.29228	4.08656	1.18642	2.55607	5.50688	.598802
1.	2.8224	4.74163	1.29615	4.09878	1.18878	2.56116	5.51785	.595238
1.66	2.8561	4.82681	1.50000	4.11096	1.19114	2.56623	5.52877	.591716
1.68	2.8900	4.91300	1.30384	4.12311	1.19348	2.57128	5.53966	.588235
1.71	2.9241	5.00021	1.30767	4.13521	1.19582	2.57631	5.55050	.584795
1.72	2.9584	5.08845	1.31149	4.14729	1.19815	2.58133	5.56130	.581395
1.73	2.9929	5.17772	1.31529	4.15933	1.20046	2.58632	5.57205	.578035
1.74	3.0276	5.26802	1.31909	4.17133	1.20277	2.59129	5.58277	.574713
1.75	3.0625	5.35938	1.32288	4.18330	1.20507	2.59625	5.59344	.571429
1.	3.0976	5.45178	1.32665	4.19524	1.20736	2.60118	5.60408	.568182
1.	3.1329	5.54523	1.33041	4.20714	1.20964	2.60610	5.61467	.564972
1.	3.1684	5.63975	1.33417	4.21900	1.21192	2.61100	5.62523	.561798
1.76	3.2041	5.73534	1.33791	4.23084	1.21418	2.61588	5.63574	.558659
1.80	3.2400	5.83200	1.34164	4.24264	1.21644	2.62074	5.64622	.555556
1.	3.2761	5.92974	1.34536	4.25441	1.21869	2.62558	5.65665	.552486
1.	3.3124	6.02857	1.34907	4.26615	1.22093	2.63041	5.66705	.549451
1.	3.3489	6.12849	1.35277	4.27785	1.22316	2.63522	5.67741	.546448
1.81	3.3856	6.22950	1.35647	4.28952	1.22539	2.64001	5.68773	.543478
1.85	3.4225	6.33163	1.36015	4.30116	1.22760	2.64479	5.69802	.540541
1.	3.4596	6.43486	1.36382	4.31277	1.22981	2.64954	5.70827	.537634
1.	3.4969	6.53920	1.36748	4.32435	1.23201	2.65428	5.71848	.534759
1.	3.5344	6.64467	1.37113	4.33590	1.23420	2.65900	5.72865	.531915
1.86	3.5721	6.75127	1.37477	4.34741	1.23639	2.66371	5.73879	.529101
1.88	3.6100	6.85900	1.37840	4.35890	1.23856	2.66840	5.74890	.526316
1.	3.6481	6.96787	1.38203	4.37035	1.24073	2.67307	5.75897	.523560
1.	3.6864	7.07789	1.38564	4.38178	1.24289	2.67773	5.76900	.520833
1.	3.7249	7.18906	1.38924	4.39318	1.24505	2.68237	5.77900	.518135
1.91	3.7636	7.30138	1.39284	4.40454	1.24719	2.68700	5.78896	.515464
1.95	3.8025	7.41488	1.39642	4.41588	1.24933	2.69161	5.79889	..512821
1.	3.8416	7.52954	1.40000	4.42719	1.25146	2.69620	5.80879	.510204
1.	3.8809	7.64537	1.40357	4.43847	1.25359	2.70078	5.81865	.507614
1.96	3.9204	7.76239	1.40712	4.44972	1.25571	2.70534	5.82848	.505051
1.99	3.9601	7.88060	1.41067	4.46094	1.25782	2.70989	5.83827	.502513
2.00	4.0000	8.00000	1.41421	4.47214	1.25992	2.71442	5.84804	.500000

n	n^2	n^3	$\sqrt{n}$	$\sqrt{10\,n}$	$\sqrt[3]{n}$	$\sqrt[3]{10\,n}$	$\sqrt[3]{100\,n}$	$\dfrac{1}{n}$
2.01	4.0401	8.12060	1.41774	4.48330	1.26202	2.71893	5.85777	.497512
2.02	4.0804	8.24241	1 42127	4.49444	1.26411	2.72343	5.86746	.495050
2.03	4.1209	8.36543	1.42478	4.50555	1.26619	2.72792	5.87713	.492611
2.04	4.1616	8.48966	1.42829	4.51664	1.26827	2.73239	5.88677	.490196
2.05	4.2025	8.61513	1.43178	4.52769	1.27033	2.73685	5.89637	.487805
2.	4.2436	8.74182	1.43527	4.53872	1.27240	2.74129	5.90594	.485437
2.	4.2849	8.86974	1.43875	4.54973	1.27445	2.74572	5.91548	.483092
2.	4.3264	8.99891	1.44222	4.56070	1.27650	2.75014	5.92499	.480769
2.06	4.3681	9.12933	1.44568	4.57165	1.27854	2.75454	5.93447	.478469
2.08	4.4100	9.26100	1.44914	4.58258	1.28058	2.75893	5.94392	.476191
2.11	4.4521	9.39393	1.45258	4.59347	1.28261	2.76330	5.95334	.473934
2.12	4.4944	9.52813	1.45602	4.60435	1.28463	2.76766	5.96273	.471698
2.13	4.5369	9.66360	1.45945	4.61519	1.28665	2.77200	5.97209	.469484
2.14	4.5796	9.80034	1.46287	4.62601	1.28866	2.77633	5.98142	.467290
2.15	4.6225	9.93838	1.46629	4.63681	1.29066	2.78065	5.99073	.465116
2.16	4.6656	10.0777	1.46969	4.64758	1.29266	2.78495	6.00000	.462963
2.17	4.7089	10.2183	1.47309	4.65833	1.29465	2.78924	6.00925	.460830
2.18	4.7524	10.3602	1.47648	4.66905	1.29664	2.79352	6.01846	.458716
2.19	4.7961	10.5035	1.47986	4.67974	1.29862	2.79779	6.02765	.456621
2.20	4.8400	10.6480	1.48324	4.69042	1.30059	2.80204	6.03681	.454546
2.	4.8841	10.7939	1.48661	4.70106	1.30256	2.80628	6.04594	.452489
2.	4.9284	10.9410	1.48997	4.71169	1.30452	2.81051	6.05505	.450451
2.	4.9729	11.0896	1.49332	4.72229	1.30648	2.81472	6.06413	.448431
2.21	5.0176	11.2394	1.49666	4.73286	1.30843	2.81892	6.07318	.446429
2.26	5.0625	11.3906	1.50000	4.74342	1.31037	2.82311	6.08220	.444444
2.	5.1076	11.5432	1.50333	4.75395	1.31231	2.82728	6.09120	.442478
2.	5.1529	11 6971	1.50665	4.76445	1.31424	2.83145	6.10017	.440529
2.	5.1984	11.8524	1.50997	4.77493	1.31617	2.83560	6.10911	.438597
2.26	5.2441	12.0090	1.51327	4.78539	1.31809	2.83974	6.11803	.436681
2.58	5.2900	12.1670	1.51658	4.79583	1.32001	2.84387	6.12693	.434783
2.	5.3361	12.3264	1.51987	4.80625	1.32192	2.84798	6.13579	.432900
2.	5.3824	12.4872	1.52315	4.81664	1.32382	2.85209	6.14463	.431035
2.	5.4289	12.6493	1.52643	4.82701	1.32572	2.85618	6 15345	.429185
2.31	5.4756	12.8129	1.52971	4.83735	1.32761	2.86026	6.16224	.427350
2.35	5.5225	12.9779	1.53297	4.84768	1.32950	2.86433	6.17101	.425532
2.36	5.5696	13.1443	1.53623	4.85798	1.33139	2 86838	6.17975	.423729
2.37	5.6169	13.3121	1.53948	4.86826	1.33326	2.87243	6.18846	.421941
2.38	5.6644	13.4813	1.54272	4.87852	1.33514	2.87646	6.19715	.420168
2.39	5.7121	13.6519	1.54596	4.88876	1.33700	2.88049	6.20582	.418410
2.40	5.7600	13.8240	1.54919	4.89898	1.33887	2.88450	6.21447	.416667
2.41	5.8081	13.9975	1.55242	4.90918	1.34072	2.88850	6.22308	.414938
2.42	5.8564	14.1725	1.55563	4.91935	1.34257	2.89249	6.23168	.413223
2.43	5.9049	14.3489	1.55885	4.92950	1.34442	2.89647	6.24025	.411523
2.44	5.9536	14.5268	1.56205	4.93964	1.34626	2.90044	6.24880	.409836
2.45	6.0025	14.7061	1.56525	4.94975	1.34810	2.90439	6.25732	.408163
2.	6.0516	14.8869	1.56844	4.95984	1.34993	2.90834	6.26583	.406504
2.	6.1009	15.0692	1.57162	4.96991	1.35176	2.91227	6.27431	.404858
2	6.1504	15.2530	1.57480	4.97996	1.35358	2.91620	6.28276	.403226
2 46	6.2001	15.4382	1.57797	4.98999	1.35540	2.92011	6.29119	.401606
2.50	6.2500	15.6250	1.58114	5.00000	1.35721	2.92402	6.29961	.400000

3

n	n^2	n^3	$\sqrt{n}$	$\sqrt{10\,n}$	$\sqrt[3]{n}$	$\sqrt[3]{10\,n}$	$\sqrt[3]{100\,n}$	$\dfrac{1}{n}$
2.51	6.3001	15.8133	1.58430	5.00999	1.35902	2.92791	6.30799	.398406
2.52	6.3504	16.0030	1.58745	5.01996	1.36082	2.93179	6.31636	.396825
2.53	6.4009	16.1943	1.59060	5.02991	1.36262	2.93567	6.32470	.395257
2 54	6.4516	16.3871	1.59374	5.03984	1.36441	2.93953	6.33303	.393701
2.55	6.5025	16.5814	1.59687	5.04975	1.36620	2.94338	6.34133	.392157
2.56	6.5536	16.7772	1.60000	5.05964	1.36798	2.94723	6.34960	.390625
2.57	6.6049	16.9746	1.60312	5.06952	1.36976	2.95106	6.35786	.389105
2.58	6.6564	17.1735	1.60624	5.07937	1.37153	2.95488	6.36610	.387597
2.59	6.7081	17.3740	1.60935	5.08920	1.37330	2.95869	6.37431	.386100
2.	6.7600	17.5760	1.61245	5.09902	1.37507	2.96250	6.38250	.384615
2.	6.8121	17.7796	1.61555	5.10882	1.37683	2.96629	6.39068	.383142
2.	6.8644	17.9847	1.61864	5.11859	1.37859	2.97007	6.39883	.381679
2.	6.9169	18.1914	1.62173	5.12835	1.38034	2.97385	6.40696	.380228
2.60	6.9696	18.3997	1.62481	5.13809	1.38208	2.97761	6.41507	.378788
2.68	7.0225	18.6096	1.62788	5.14782	1.38383	2.98137	6.42316	.377359
2.	7.0756	18.8211	1.63095	5.15752	1.38557	2.98511	6.43123	.375940
2.	7.1289	19.0342	1.63401	5.16720	1.38730	2.98885	6.43928	.374532
2.	7.1824	19.2488	1.63707	5.17687	1.38903	2.99257	6.44731	.373134
2.66	7.2361	19.4651	1.64012	5.18652	1.39076	2.99629	6.45531	.371747
2.68	7.2900	19.6830	1.64317	5.19615	1.39248	3.00000	6.46330	.370370
2.71	7.3441	19.9025	1.64621	5.20577	1.39419	3.00370	6.47127	.369004
2.??	7.3984	20.1236	1.64924	5.21536	1.39591	3.00739	6.47922	.367647
2.	7.4529	20.3464	1.65227	5.22494	1.39761	3.01107	6.48715	.366300
2.72	7.5076	20.5708	1.65529	5.23450	1.39932	3 01474	6.49507	.364964
2.73	7.5625	20.7969	1.65831	5.24404	1.40102	3.01841	6.50296	.363636
2.	7.6176	21.0246	1.66132	5.25357	1.40272	3.02206	6.51083	.362319
2.	7.6729	21.2539	1.66433	5.26308	1.40441	3.02571	6.51868	.361011
2.	7.7284	21.4850	1.66733	5.27257	1.40610	3.02934	6.52652	.359712
2.76	7.7841	21.7176	1.67033	5.28205	1.40778	3.03297	6.53434	.358423
2.88	7.8400	21.9520	1.67332	5.29150	1.40946	3.03659	6.54213	.357142
2.	7.8961	22.1880	1.67631	5.30094	1.41114	3.04020	6.54991	.355872
2.	7.9524	22.4258	1.67929	5.31037	1.41281	3.04380	6.55767	.354610
2.	8.0089	22.6652	1.68226	5.31977	1.41448	3.04740	6.56541	.353357
2.	8.0656	22 9063	1.68523	5.32917	1.41614	3.05098	6.57314	.352113
2.81	8.1225	23.1491	1.68819	5.33854	1.41780	3.05456	6.58084	.350877
2	8.1796	23.3937	1.69115	5.34790	1.41946	3.05813	6.58853	.349650
2	8.2369	23.6399	1 69411	5.35724	1.42111	3.06169	6.59620	.348432
2	8.2944	23.8879	1.69706	5.36656	1.42276	3.06524	6.60385	.347222
2 86	8.3521	24.1376	1.70000	5.37587	1.42440	3.06878	6.61149	.346021
2.88	8.4100	24.3890	1.70294	5.38516	1.42604	3.07232	6.61911	.344828
2.	8.4681	24.6422	1.70587	5.39444	1.42768	3.07585	6.62671	.343643
2.	8.5264	24.8971	1.70880	5.40370	1.42931	3.07936	6.63429	.342466
2.	8.5849	25.1538	1.71172	5.41295	1.43094	3.08287	6.64185	.341297
2.91	8.6436	25.4122	1.71464	5.42218	1.43257	3.08638	6.64940	.340136
2.98	8.7025	25.6724	1.71756	5.43139	1.43419	3.08987	6.65693	.338985
96	8.7616	25.9343	1.72047	5.44059	1.43581	3.09336	6.66444	.337838
2.97	8.8209	26.1981	1.72337	5.44977	1.43743	3.09684	6.67194	.336700
98	8.8804	26.4636	1.72627	5.45894	1.43904	3.10031	6.67942	.335571
99	8.9401	26.7309	1.72916	5.46809	1.44065	3.10378	6.68688	.334448
3.00	9.0000	27.0000	1.73205	5 47723	1.44225	3.10723	6.69433	.333333

n	n^2	n^3	$\sqrt{n}$	$\sqrt{10\,n}$	$\sqrt[3]{n}$	$\sqrt[3]{10\,n}$	$\sqrt[3]{100\,n}$	$\dfrac{1}{n}$
3 01	9.0601	27.2709	1.73494	5.48635	1.44385	3.11068	6.70176	.332226
3 02	9.1204	27.5436	1.73781	5.49545	1.44545	3.11412	6.70917	.331126
3.03	9.1809	27.8181	1.74069	5.50454	1.44704	3.11755	6.71657	.330033
3.04	9.2416	28.0945	1.74356	5.51362	1.44863	3.12098	6.72395	.328947
3.05	9.3025	28.3726	1.74642	5.52268	1 45022	3.12440	6.73132	.327869
3 06	9.3636	28.6526	1.74929	5.53173	1.45180	3.12781	6.73866	.326797
.07	9.4249	28.9344	1.75214	5.54076	1.45338	3.13121	6.74600	.325733
8 08	9.4864	29.2181	1.75499	5.54977	1.45496	3.13461	6.75331	.324675
3.09	9.5481	29.5036	1.75784	5.55878	1.45653	3.13800	6.76061	.323625
3.10	9.6100	29.7910	1.76068	5.56776	1.45810	3.14138	6.76790	.322581
.11	9.6721	30.0802	1.76352	5.57674	1.45967	3.14475	6.77517	.321543
.12	9.7344	30.3713	1.76635	5.58570	1.46123	3.14812	6.78242	.320513
.13	9.7969	30.6643	1.76918	5.59464	1.46279	3.15148	6.78966	.319489
.14	9.8596	30.9591	1.77200	5.60357	1.46434	3.15484	6.79688	.318471
8.15	9.9225	31.2559	1.77482	5.61249	1.46590	3.15818	6.80409	.317460
3.16	9.9856	31.5545	1.77764	5.62139	1.46745	3.16152	6.81 28	.316456
8.17	10.0489	31.8550	1.78045	5 63028	1.46899	3.16485	6.81846	.315457
3.18	10.1124	32.1574	1.78326	5.63915	1.47054	3.16817	6.82562	.314465
3.19	10.1761	32.4618	1.78606	5.64801	1.47208	3.17149	6.83277	.313480
3.20	10.2400	32.7680	1.78885	5.65685	1.47361	3.17480	6.83990	.312500
.21	10.3041	33.0762	1.79165	5.66569	1.47515	3.17811	6.84702	.311527
.22	10.3684	33.3862	1.79444	5.67450	1.47668	3.18140	6.85412	.310559
23	10.4329	33.6983	1.79722	5.68331	1.47820	3.18469	6.86121	.309598
24	10.4976	34.0122	1.80000	5.69210	1.47973	3.18798	6.86829	.308642
8.25	10.5625	34.3281	1.80278	5.70088	1.48125	3.19125	6.87534	.307692
8.	10.6276	34.6460	1.80555	5.70964	1.48277	3.19452	6 88239	.306749
8.	10.6929	34.9658	1.80831	5.71839	1.48428	3.19779	6.88942	.305810
3.	10.7584	35.2876	1.81108	5.72713	1.48579	3.20104	6 89643	.304878
3.26	10.8241	35.6129	1.81384	5.73585	1.48730	3.20429	6 90344	.303951
3.88	10.8900	35.9370	1.81659	5.74456	1.48881	3.20753	6.91042	.303030
3.	10.9561	36.2647	1.81934	5.75326	1.49031	3.21077	6 91740	.302115
3.	11.0224	36.5944	1.82209	5.76194	1.49181	3.21400	6.92436	.301205
3.	11.0889	36.9260	1.82483	5.77062	1.49330	3.21723	6 93130	.300300
3.31	11.1556	37.2597	1.82757	5.77927	1.49480	3.22044	6.93823	.299401
3.88	11.2225	37.5954	1.83030	5.78792	1.49629	3.22365	6.94515	.298508
8	11.2896	37.9331	1.83303	5.79655	1.49777	3.22686	6.95205	.297619
3.	11.3569	38.2728	1.83576	5.80517	1.49926	3.23005	6.95894	.296736
8.	11.4244	38.6145	1.83848	5.81378	1.50074	3.23325	6.96582	.295858
	11.4921	38.9582	1.84120	5.82237	1.50222	3.23643	6 97268	.294985
3.36	11.5600	39.3040	1.84391	5.83095	1.50369	3.23961	6.97953	.294118
3.41	11.6281	39.6518	1.84662	5.83952	1.50517	3.24278	6 98637	.293255
3.42	11.6964	40 0017	1.84932	5.84808	1.50664	3.24595	6.99319	.292398
8.43	11.7649	40.3536	1.85203	5.85662	1.50810	3.24911	7 00000	.291545
44	11.8336	40.7076	1.85472	5.86515	1.50957	3.25227	7.00680	.290698
8.45	11.9025	41.0636	1.85742	5.87367	1.51103	3.25542	7.01358	289855
8.46	11.9716	41.4217	1.86011	5.88218	1.51249	3.25856	7.02035	.289017
3.47	12.0409	41.7819	1.86279	5.89067	1.51394	3.26169	7.02711	.288184
3.48	12.1104	42.1442	1.86548	5.89915	1.51540	3.26482	7.03385	.287356
3.49	12.1801	42.5085	1.86815	5.90762	1.51685	3.26795	7.04058	.286533
3.50	12.2500	42.8750	1.87083	5.91608	1.51829	3.27107	7.04730	.285714

n	n^2	n^3	$\sqrt{n}$	$\sqrt{10\,n}$	$\sqrt[3]{n}$	$\sqrt[3]{10\,n}$	$\sqrt[3]{100\,n}$	$\dfrac{1}{n}$
3.51	12.3201	43.2436	1.87350	5.92453	1.51974	3.27418	7.05400	.284900
3.52	12.3904	43.6142	1.87617	5.93296	1.52118	3.27729	7.06070	.284091
3.53	12.4609	43.9870	1.87883	5.94138	1.52262	3.28039	7.06738	.283286
3.54	12.5316	44.3619	1.88149	5.94979	1.52406	3.28348	7.07404	.282486
3.55	12.6025	44.7389	1.88414	5.95819	1.52549	3.28657	7.08070	.281690
3.56	12.6736	45.1180	1.88680	5.96657	1.52692	3.28965	7.08734	.280899
3.57	12.7449	45.4993	1.88944	5.97495	1.52835	3.29273	7.09397	.280112
3.58	12.8164	45.8827	1.89209	5.98331	1.52978	3.29580	7.10059	.279330
3 59	12.8881	46.2683	1.89473	5.99166	1.53120	3.29887	7.10719	.278552
3.60	12.9600	46.6560	1.89737	6.00000	1.53262	3.30193	7.11379	.277778
3.61	13.0321	47.0459	1.90000	6.00833	1.53404	3.30498	7.12037	.277008
3.62	13.1044	47.4379	1.90263	6.01664	1.53545	3.30803	7.12694	.276243
3.63	13.1769	47.8321	1.90526	6.02495	1.53686	3.31107	7.13349	.275482
3.64	13.2496	48.2285	1.90788	6.03324	1.53827	3.31411	7.14004	.274725
3.65	13.3225	48.6271	1.91050	6.04152	1.53968	3.31714	7.14657	.273973
3.	13.3956	49.0279	1.91311	6.04979	1.54109	3.32017	7.15309	.273224
3.	13.4689	49.4309	1.91572	6.05805	1.54249	3.32319	7.15960	.272480
3.	13.5424	49.8360	1.91833	6.06630	1.54389	3.32621	7.16610	.271739
3.66	13.6161	50.2434	1.92094	6.07454	1.54529	3.32922	7.17258	.271003
3.66	13.6900	50.6530	1.92354	6.08276	1.54668	3.33222	7.17905	.270270
3.71	13.7641	51.0648	1.92614	6.09098	1.54807	3.33522	7.18552	.269542
3.72	13.8384	51.4788	1.92873	6.09918	1.54946	3.33822	7.19197	.268817
3.73	13.9129	51.8951	1.93132	6.10737	1.55085	3.34120	7.19841	.268097
3.74	13.9876	52.3136	1.93391	6.11555	1.55223	3.34419	7.20483	.267380
3.75	14.0625	52.7344	1.93649	6.12372	1.55362	3.34716	7.21125	.266667
3.	14.1376	53.1574	1.93907	6.13188	1.55500	3.35014	7.21765	.265957
3.	14.2129	53.5826	1.94165	6.14003	1.55637	3.35310	7.22405	.265252
3.	14.2884	54.0102	1.94422	6.14817	1.55775	3.35607	7.23043	.264550
3.76	14.3641	54.4399	1.94679	6.15630	1.55912	3.35902	7.23680	.263852
3.80	14.4400	54.8720	1.94936	6.16441	1.56049	3.36198	7.24316	.263158
3.81	14 5161	55.3063	1.95192	6.17252	1.56186	3.36492	7.24950	.262467
3.82	14.5924	55.7430	1.95448	6.18061	1.56322	3.36786	7.25584	.261780
3.83	14.6689	56.1819	1.95704	6.18870	1.56459	3.37080	7.26217	.261097
3.84	14.7456	56.6231	1.95959	6.19677	1.56595	3.37373	7.26848	.260417
3.85	14.8225	57.0666	1.96214	6.20484	1.56731	3.37666	7.27479	.259740
3.86	14.8996	57.5125	1.96469	6.21289	1.56866	3.37958	7.28108	.259067
3 87	14.9769	57.9606	1.96723	6.22093	1.57001	3.38249	7.28736	.258398
3.88	15.0544	58.4111	1.96977	6.22896	1.57137	3.38540	7.29363	.257732
3 89	15.1321	58 8639	1.97231	6.23699	1.57271	3.38831	7.29989	.257069
3.90	15.2100	59.3190	1.97484	6.24500	1.57406	3.39121	7.30614	.256410
3.91	15.2881	59.7765	1.97737	6.25300	1.57541	3.39411	7.31238	.255755
3 92	15.3664	60.2363	1.97990	6.26099	1.57675	3.39700	7.31861	.255102
3.93	15.4449	60.6985	1.98242	6.26897	1.57809	3.39988	7.32483	.254453
3.94	15.5236	61.1630	1.98494	6.27694	1.57942	3.40277	7.33104	.253807
3.95	15.6025	61.6299	1.98746	6.28490	1.58076	3.40564	7.33723	.253165
3.	15.6816	62.0991	1.98997	6.29285	1.58209	3.40851	7.34342	.252525
	15.7609	62.5708	1.99249	6.30079	1.58342	3.41138	7.34960	.251889
	15.8404	63.0448	1 99499	6.30872	1.58475	3.41424	7.35576	.251256
2.96	15.9201	63.5212	1.99750	6.31664	1.58608	3.41710	7.36192	.250627
2.00	16.0000	64.0000	2.00000	6.32456	1.58740	3.41995	7.36806	.250000

n	n^2	n^3	$\sqrt{n}$	$\sqrt{10\,n}$	$\sqrt[3]{n}$	$\sqrt[3]{10\,n}$	$\sqrt[3]{100\,n}$	$\dfrac{1}{n}$
4.01	16.0801	64.4812	2.00250	6.33246	1.58872	3.42280	7.37420	.249377
4.02	16.1604	64.9648	2.00499	6.34035	1.59004	3.42564	7.38032	.248756
4.03	16.2409	65.4508	2.00749	6.34823	1.59136	3.42848	7.38644	.248139
4.04	16.3216	65.9393	2.00998	6.35610	1.59267	3.43131	7.39254	.247525
4.05	16.4025	66.4301	2.01246	6.36396	1.59399	3.43414	7.39864	.246914
4.06	16.4836	66.9234	2.01494	6.37181	1.59530	3.43697	7.40472	.246305
4.07	16.5649	67.4191	2.01742	6.37966	1.59661	3.43979	7.41080	.245700
4.08	16.6464	67.9173	2.01990	6.38749	1.59791	3.44260	7.41686	.245098
4.09	16.7281	68.4179	2.02237	6.39531	1.59922	3.44541	7.42291	.244499
4.10	16.8100	68.9210	2.02485	6.40312	1.60052	3.44822	7.42896	.243902
4.11	16.8921	69.4265	2.02731	6.41093	1.60182	3.45102	7.43499	.243309
4.12	16.9744	69.9345	2.02978	6.41872	1.60312	3.45382	7.44102	.242718
4.13	17.0569	70.4450	2.03224	6.42651	1.60441	3.45661	7.44703	.242131
4.14	17.1396	70.9579	2.03470	6.43428	1.60571	3.45939	7.45304	.241546
4.15	17.2225	71.4734	2.03715	6.44205	1.60700	3.46218	7.45904	.240964
4.16	17.3056	71.9913	2.03961	6.44981	1.60829	3.46496	7.46502	.240385
4.17	17.3889	72.5117	2.04206	6.45755	1.60958	3.46773	7.47100	.239808
4.18	17.4724	73.0346	2.04450	6.46529	1.61086	3.47050	7.47697	.239234
4.19	17.5561	73.5601	2.04695	6.47302	1.61215	3.47327	7.48292	.238664
4.20	17.6400	74.0880	2.04939	6.48074	1.61343	3.47603	7.48887	.238095
4.21	17.7241	74.6185	2.05183	6.48845	1.61471	3.47878	7.49481	.237530
4.22	17.8084	75.1514	2.05426	6.49615	1.61599	3.48154	7.50074	.236967
4.23	17.8929	75.6870	2.05670	6.50385	1.61726	3.48428	7.50666	.236407
4.24	17.9776	76.2250	2.05913	6.51153	1.61853	3.48703	7.51257	.235849
4.25	18.0625	76.7656	2.06155	6.51920	1.61981	3.48977	7.51847	.235294
4.26	18.1476	77.3088	2.06398	6.52687	1.62108	3.49250	7.52437	.234742
4.27	18.2329	77.8545	2.06640	6.53452	1.62234	3.49523	7.53025	.234192
4.28	18.3184	78.4028	2.06882	6.54217	1.62361	3.49796	7.53612	.233645
4.29	18.4041	78.9536	2.07123	6.54981	1.62487	3.50068	7.54199	.233100
4.30	18.4900	79.5070	2.07364	6.55744	1.62613	3.50340	7.54784	.232558
4.	18.5761	80.0630	2.07605	6.56506	1.62739	3.50611	7.55369	.232019
4.	18.6624	80.6216	2.07846	6.57267	1.62865	3.50882	7.55953	.231482
4.	18.7489	81.1827	2.08087	6.58027	1.62991	3.51153	7.56535	.230947
4.31	18.8356	81.7465	2.08327	6.58787	1.63116	3.51423	7.57117	.230415
4.35	18.9225	82.3129	2.08567	6.59545	1.63241	3.51692	7.57698	.229885
4.36	19.0096	82.8819	2.08806	6.60303	1.63366	3.51962	7.58279	.229358
4.37	19.0969	83.4535	2.09045	6.61060	1.63491	3.52231	7.58858	.228833
4.38	19.1844	84.0277	2.09284	6.61816	1.63616	3.52499	7.59436	.228311
4.39	19.2721	84.6045	2.09523	6.62571	1.63740	3.52767	7.60014	.227790
4.40	19.3600	85.1840	2.09762	6.63325	1.63864	3.53035	7.60590	.227273
4.41	19.4481	85.7661	2.10000	6.64078	1.63988	3.53302	7.61166	.226757
4.42	19.5364	86.3509	2.10238	6.64831	1.64112	3.53569	7.61741	.226244
4.43	19.6249	86.9383	2.10476	6.65582	1.64236	3.53835	7.62315	.225734
4.44	19.7136	87.5284	2.10713	6.66333	1.64359	3.54101	7.62888	.225225
4.45	19.8025	88.1211	2.10950	6.67083	1.64483	3.54367	7.63461	.224719
4.	19.8916	88.7165	2.11187	6.67832	1.64606	3.54632	7.64032	.224215
4.	19.9809	89.3146	2.11424	6.68581	1.64729	3.54897	7.64603	.223714
4.	20.0704	89.9154	2.11660	6.69328	1.64851	3.55162	7.65172	.223214
4.46	20.1601	90.5188	2.11896	6.70075	1.64974	3.55426	7.65741	.222717
4.49	20.2500	91.1250	2.12132	6.70820	1.65096	3.55689	7.66309	.222222

n	n^2	n^3	$\sqrt{n}$	$\sqrt{10\,n}$	$\sqrt[3]{n}$	$\sqrt[3]{10\,n}$	$\sqrt[3]{100\,n}$	$\dfrac{1}{n}$
4.51	20.3401	91.7339	2.12368	6.71565	1.65219	3.55953	7.66877	.221730
4.52	20.4304	92.3454	2.12603	6.72309	1.65341	3.56215	7.67443	.221239
4.53	20.5209	92.9597	2.12838	6.73053	1.65462	3.56478	7.68009	.220751
4.54	20.6116	93.5767	2.13073	6.73795	1.65584	3.56740	7.68573	.220264
4.55	20.7025	94.1964	2.13307	6.74537	1.65706	3.57002	7.69137	.219780
4.56	20.7936	94.8188	2.13542	6.75278	1.65827	3.57263	7 69700	.219298
4.57	20.8849	95.4440	2.13776	6.76018	1.65948	3.57524	7.70262	.218818
4.58	20.9764	96 0719	2.14009	6.76757	1.66069	3.57785	7.70824	.218341
4.59	21.0681	96.7026	2.14243	6.77495	1.66190	3.58045	7.71384	.217865
4.60	21.1600	97.3360	2.14476	6.78233	1.66310	3.58305	7.71944	.217391
4.	21.2521	97 9722	2.14709	6.78970	1 66431	3.58564	7.72503	.216920
4.	21.3444	98.6111	2.14942	6.79706	1.66551	3.58823	7.73061	.216450
4.	21.4369	99.2528	2.15174	6.80441	1.66671	3.59082	7.73619	.215983
4.61	21.5296	99.8973	2.15407	6.81175	1.66791	3.59340	7.74175	.215517
4.63	21.6225	100.545	2.15639	6.81909	1.66911	3.59598	7.74731	.215054
4.	21.7156	101.195	2.15870	6.82642	1.67030	3.59856	7.75286	.214592
4.	21.8089	101.848	2.16102	6.83374	1.67150	3.60113	7.75840	.214133
4.	21.9024	102.503	2.16333	6.84105	1.67269	3.60370	7.76394	.213675
4.66	21.9961	103.162	2.16564	6.84836	1.67388	3.60626	7.76946	.213220
4.69	22.0900	103.823	2.16795	6.85565	1.67507	3.60883	7.77498	.212766
4.71	22.1841	104.487	2.17025	6.86294	1.67626	3.61138	7.78049	.212314
4.72	22.2784	105.154	2.17256	6.87023	1.67744	3.61394	7.78599	.211864
4.73	22.3729	105.824	2.17486	6.87750	1.67863	3.61649	7.79149	.211417
4.74	22.4676	106.496	2.17715	6.88477	1.67981	3 61904	7.79697	.210971
4.75	22.5625	107.172	2.17945	6.89202	1.68099	3.62158	7.80245	.210526
4.76	22.6576	107.850	2.18174	6.89928	1.68217	3.62412	7.80793	.210084
4.77	22.7529	108.531	2.18403	6.90652	1.68334	3.62665	7.81339	.209644
4.78	22.8484	109.215	2.18632	6.91375	1.68452	3.62919	7.81885	.209205
4.79	22.9441	109.902	2.18861	6.92098	1.68569	3.63171	7.82429	.208768
4.80	23.0400	110.592	2.19089	6.92820	1.68687	3.63424	7.82974	.208333
4.	23.1361	111.285	2.19317	6.93542	1.68804	3.63676	7.83517	.207900
4.	23.2324	111.980	2.19545	6.94262	1.68920	3.63928	7.84059	.207469
4.	23.3289	112.679	2.19773	6.94982	1.69037	3.64180	7.84601	.207039
4.	23.4256	113.380	2.20000	6.95701	1.69154	3.64431	7.85142	.206612
4.84	23.5225	114.084	2.20227	6.96419	1.69270	3.64682	7.85683	.206186
4.86	23.6196	114.791	2.20454	6.97137	1.69386	3.64932	7.86222	.205761
4.87	23.7169	115.501	2.20681	6.97854	1.69503	3.65182	7.86761	.205339
4.88	23.8144	116.214	2.20907	6.98570	1.69619	3.65432	7.87299	.204918
4.89	23.9121	116.930	2.21133	6.99285	1.69734	3.65682	7.87837	.204499
4.90	24 0100	117.649	2.21359	7.00000	1.69850	3.65931	7.88374	.204082
4.	24.1081	118.371	2.21585	7.00714	1.69965	3.66179	7.88909	.203666
4.	24.2064	119.095	2.21811	7.01427	1.70081	3 66428	7.89445	.203252
4.	24.3049	119.823	2.22036	7.02140	1.70196	3.66676	7.89979	.202840
4.91	24.4036	120.554	2 22261	7.02851	1.70311	3.66924	7.90513	.202429
4.93	24.5025	121.287	2.22486	7.03562	1.70426	3.67171	7.91046	.202020
4.	24.6016	122.024	2.22711	7.04273	1.70540	3.67418	7.91578	.201613
4.	24.7009	122.763	2.22935	7.04982	1.70655	3.67665	7.92110	.201207
4.	24.8004	123.506	2.23159	7.05691	1.70769	3.67911	7.92641	.200803
4.96	24.9001	124.251	2.23383	7.06399	1.70884	3.68157	7.93171	.200401
5.00	25.0000	125.000	2.23607	7.07107	1.70998	3.68403	7.93701	.200000

n	n^2	n^3	$\sqrt{n}$	$\sqrt{10\,n}$	$\sqrt[3]{n}$	$\sqrt[3]{10\,n}$	$\sqrt[3]{100\,n}$	$\dfrac{1}{n}$
5.01	25.1001	125.752	2.23830	7.07814	1.71112	3.68649	7.94229	.199601
.02	25.2004	126.506	2.24054	7.08520	1.71225	3.68894	7.94757	.199203
8.03	25.3009	127.264	2.24277	7.09225	1.71339	3.69138	7.95285	.198807
5.04	25.4016	128.024	2.24499	7.09930	1.71452	3.69383	7.95811	.198413
.05	25.5025	128.788	2.24722	7.10634	1.71566	3.69627	7.96337	.198020
5.06	25.6036	129.554	2.24944	7.11337	1.71679	3.69871	7.96863	.197629
5.07	25.7049	130.324	2.25167	7.12039	1.71792	3.70114	7.97387	.197239
5.08	25.8064	131.097	2.25389	7.12741	1.71905	3.70358	7.97911	.196850
5.09	25.9081	131.872	2.25610	7.13442	1.72017	3.70600	7.98434	.196464
5.10	26.0100	132.651	2.25832	7.14143	1.72130	3.70843	7.98957	.196078
5.11	26.1121	133.433	2.26053	7.14843	1.72242	3.71085	7.99479	.195695
5.12	26.2144	134.218	2.26274	7.15542	1.72355	3.71327	8.00000	.195313
5.13	26.3169	135.006	2.26495	7.16240	1.72467	3.71566	8.00520	.194932
5.14	26.4196	135.797	2.26716	7.16938	1.72579	3.71816	8.01040	.194553
5.15	26.5225	136.591	2.26936	7.17635	1.72691	3.72051	8.01559	.194175
5.16	26.6256	137.388	2.27156	7.18331	1.72802	3.72292	8.02078	.193798
5.17	26.7289	138.188	2.27376	7.19027	1.72914	3.72532	8.02596	.193424
5.18	26.8324	138.992	2.27596	7.19722	1.73025	3.72772	8.03113	.193050
5.19	26.9361	139.798	2.27816	7.20417	1.73137	3.73012	8.03629	.192678
5.20	27.0400	140.608	2.28035	7.21110	1.73248	3.73251	8.04145	.192308
5. 1	27.1441	141.421	2.28254	7.21803	1.73359	3.73490	8.04660	.191939
5.22	27.2484	142.237	2.28473	7.22496	1.73470	3.73729	8.05175	.191571
5.23	27.3529	143.056	2.28692	7.23187	1.73580	3.73968	8.05689	.191205
5.24	27.4576	143.878	2.28910	7.23878	1.73691	3.74206	8.06202	.190840
5.25	27.5625	144.703	2.29129	7.24569	1.73801	3.74443	8.06714	.190476
5.26	27.6676	145.532	2.29347	7.25259	1.73912	3.74681	8.07226	.190114
5.27	27.7729	146.363	2.29565	7.25948	1.74022	3.74918	8.07737	.189753
5.28	27.8784	147.198	2.29783	7.26636	1.74132	3.75158	8.08248	.189394
5.29	27.9841	148.036	2.30000	7.27324	1.74242	3.75392	8.08758	.189036
5.30	28.0900	148.877	2.30217	7.28011	1.74351	3.75629	8.09267	.188679
	28.1961	149.721	2.30434	7.28697	1.74461	3.75865	8.09776	.188324
	28.3024	150.569	2.30651	7.29383	1.74570	3.76100	8.10284	.187970
	28.4089	151.419	2.30868	7.30068	1.74680	3.76336	8.10791	.187617
5.31	28.5156	152.273	2.31084	7.30753	1.74789	3.76571	8.11298	.187266
5.3	28.6225	153.130	2.31301	7.31437	1.74898	3.76806	8.11804	.186916
5.36	28.7296	153.991	2.31517	7.32120	1.75007	3.77041	8.12310	.186567
5.37	28.8369	154.854	2.31733	7.32803	1.75116	3.77275	8.12814	.186220
5.38	28.9444	155.721	2.31948	7.33485	1.75224	3.77509	8.13319	.185874
5.39	29.0521	156.591	2.32164	7.34166	1.75333	3.77740	8.13822	.185529
5.40	29.1600	157.464	2.32379	7.34847	1.75441	3.77976	8.14325	.185185
5.41	29.2681	158.340	2.32594	7.35527	1.75549	3.78210	8.14828	.184843
42	29.3764	159.220	2.32809	7.36206	1.75657	3.78442	8.15329	.184502
43	29.4849	160.103	2.33024	7.36885	1.75765	3.78675	8.15831	.184162
5.44	29.5936	160.989	2.33238	7.37564	1.75873	3.78907	8.16331	.183824
5.45	29.7025	161.879	2.33452	7.38241	1.75981	3.79139	8.16831	.183486
5.	29.8116	162.771	2.33666	7.38918	1.76088	3.79371	8.17330	.183150
	29.9209	163.667	2.33880	7.39594	1.76196	3.79603	8.17829	.182815
	30.0304	164.567	2.34094	7.40270	1.76303	3.79834	8.18327	.182482
	30.1401	165.469	2.34307	7.40945	1.76410	3.80065	8.18824	.182149
5.16	30.2500	166.375	2.34521	7.41620	1.76517	3.80295	8.19321	.181818

n	n^2	n^3	$\sqrt{n}$	$\sqrt{10\,n}$	$\sqrt[3]{n}$	$\sqrt[3]{10\,n}$	$\sqrt[3]{100\,n}$	$\dfrac{1}{n}$
5.51	30.3601	167.284	2.34734	7.42294	1.76624	3.80526	8.19818	.181488
5.52	30.4704	168.197	2.34947	7.42967	1.76731	3 80756	8.20313	.181159
5.53	30.5809	169.112	2.35160	7.436+0	1.76838	3.80986	8.20808	.180832
5.54	30.6916	170.031	2.35372	7.44312	1.76944	3.80115	8.21303	.180505
5.55	30.8025	170.954	2.35584	7.44983	1.77051	3.81444	8.21797	.180180
5.	30.9136	171.880	2.35797	7.45654	1.77157	3.81673	8.22290	.179856
5.	31.0249	172.809	2.36008	7.46324	1.77263	3.81902	8.22783	.179533
5.	31.1364	173.741	2.36220	7.46994	1.77369	3.82130	8.23275	.179212
5.56	31.2481	174.677	2.36432	7.47663	1.77475	3.82358	8.23766	.178891
5.59	31.3600	175.616	2.36643	7.48331	1.77581	3.82586	8.24257	.178571
5.	31.4721	176.558	2.36854	7.48999	1.77686	3.82814	8.24747	.178253
5.	31.5844	177.504	2.37065	7.49667	1.77792	3.83041	8.25237	.177936
5.	31.6969	178.454	2.37276	7.50333	1.77897	3.83268	8.25726	.177620
5.61	31.8096	179.406	2.37487	7.50999	1.78003	3.83495	8.26215	.177305
5.65	31.9225	180.362	2.37697	7.51665	1.78108	3.83721	8.26703	.176991
5.66	32.0356	181.321	2.37908	7.52330	1.78213	3.83948	8.27190	.176678
5.67	32.1489	182.284	2.38118	7.52994	1.78318	3.84174	8.27677	.176367
5.68	32.2624	183.250	2.38328	7.53658	1.78422	3.84400	8.28164	.176056
5.69	32.3761	184.220	2.38537	7.54321	1.78527	3.84625	8.28649	.175747
5.70	32.4900	185.193	2.38747	7.54983	1.78632	3.84850	8.29134	.175439
5.71	32.6041	186.169	2.38956	7.55645	1.78736	3.85075	8.29619	.175131
5.72	32.7184	187.149	2.39165	7.56307	1.78840	3.85300	8.30103	.174825
5.73	32.8329	188.133	2.39374	7.56968	1.78944	3.85524	8.30587	.174520
5.74	32.9476	189.119	2.39583	7.57628	1.79048	3.85748	8.31069	.174216
5.75	33.0625	190.109	2.39792	7.58288	1.79152	3.85972	8.31552	.173913
5.	33.1776	191.103	2.40000	7.58947	1.79256	3.86196	8.32034	.173611
5.	33.2929	192.100	2.40208	7.59605	1.79360	3.86419	8.32515	.173310
5.	33.4084	193.101	2.40416	7.60263	1.79463	3.86642	8.32995	.173010
5.76	33.5241	194.105	2.40624	7.60920	1.79567	3.86865	8.33476	.172712
5.80	33.6400	195.112	2.40832	7.61577	1.79670	3.87088	8.33955	.172414
5.	33.7561	196.123	2.41039	7.62234	1.79773	3.87310	8.34434	.172117
5.	33.8724	197.137	2.41247	7.62389	1.79876	3.87532	8.34913	.171821
5.	33.9889	198.155	2.41454	7.63544	1.79979	3.87754	8.35390	.171527
5.81	34.1056	199.177	2.41661	7.64199	1.80082	3.87975	8.35868	.171233
5.85	34.2225	200.202	2.41868	7.64853	1.80185	3.88197	8.36345	.170940
86	34.3396	201.230	2.42074	7.65506	1.80288	3.88418	8.36821	.170649
87	34.4569	202.262	2.42281	7.66159	1.80390	3.88639	8.37297	.170358
88	34.5744	203.297	2.42487	7.66812	1.80492	3.88859	8.37772	.170068
.89	34.6921	204.336	2.42693	7.67463	1.80595	3.89082	8.38247	.169779
5.90	34.8100	205.379	2.42899	7.68115	1.80697	3.89300	8.38721	.169492
5.	34.9281	206.425	2.43105	7.68765	1.80799	3.89520	8.39194	.169205
	35.0464	207.475	2.43311	7.69415	1.80901	3.89739	8.39667	.168919
5.	35.1649	208.528	2.43516	7.70065	1.81003	3.89958	8.40140	.168634
5.91	35.2836	209.585	2.43721	7.70714	1.81104	3.90177	8.40612	.168350
5.95	35.4025	210.645	2.43926	7.71362	1.81206	3.90396	8.41083	.168067
5.	35.5216	211.709	2.44131	7.72010	1.81307	3.90615	8.41554	.167785
5.	35.6409	212.776	2.44336	7.72658	1.81409	3.90833	8.42025	.167504
5.96	35.7604	213.847	2.44540	7.73305	1.81510	3.91051	8.42494	.167224
5.98	35.8801	214.922	2.44745	7.73951	1.81611	3.91269	8.42964	.166945
6.00	36.0000	216.000	2.44949	7.74597	1.81712	3.91487	8.43433	.166667

n	n^2	n^3	$\sqrt{n}$	$\sqrt{10\,n}$	$\sqrt[3]{n}$	$\sqrt[3]{10\,n}$	$\sqrt[3]{100\,n}$	$\dfrac{1}{n}$
.	36.1201	217.082	2.45153	7.75242	1.81813	3.91704	8.43901	.166389
	36.2404	218.167	2.45357	7.75887	1.81914	3 91921	8.44369	.166113
	36.3609	219.256	2.45561	7.76531	1.82014	3.92138	8.44836	.165838
6.01	36.4816	220.349	2.45764	7.77174	1.82115	3.92355	8.45303	.165568
6.02	36.6025	221.445	2.45967	7.77817	1.82215	3.92571	8.45769	.165289
.	36.7236	222.545	2.46171	7.78460	1.82316	3.92787	8.46235	.165017
	36 8449	223.649	2.46374	7.79102	1.82416	3.93003	8.46700	.164745
.	36.9664	224.756	2.46577	7.79744	1.82516	3.93219	8.47165	.164474
6.06	37.0881	225.867	2.46779	7.80385	1.82616	3.93434	8.47629	.164204
6.09	37.2100	226.981	2.46982	7.81025	1.82716	3.93650	8.48093	.163934
11	37.3321	228.099	2.47184	7.81665	1.82816	3.93865	8.48556	.163666
12	37.4544	229.221	2.47386	7.82304	1.82915	3.94079	8.49018	.163399
13	37.5769	230.346	2.47588	7.82943	1.83015	3.94294	8.49481	.163132
6.14	37.6996	231.476	2.47790	7.83582	1.83115	3.94508	8.49942	.162866
6.15	37.8225	232.608	2.47992	7.84219	1.83214	3.94722	8.50404	.162602
16	37.9456	233.745	2.48193	7.84857	1.83313	3.94936	8.50864	.162338
17	38.0689	234.885	2.48395	7.85493	1.83412	3.95150	8.51324	.162075
18	38.1924	236.029	2.48596	7.86130	1.83511	3.95363	8.51784	.161812
6.19	38.3161	237.177	2.48797	7.86766	1.83610	3.95576	8.52243	.161551
6.20	38.4400	238.328	2.48998	7.87401	1.83709	3.95789	8.52702	.161290
6.21	38.5641	239.483	2.49199	7.88036	1.83808	3.96002	8.53160	.161031
22	38.6884	240 642	2.49399	7.88670	1.83906	3.96214	8.53618	.160772
23	38 8129	241.804	2.49600	7.89303	1.84005	3.96426	8.54075	.160514
6.24	38.9376	242.971	2.49800	7.89937	1.84103	3.96639	8.54532	.160256
6.25	39.0625	244.141	2.50000	7.90569	1.84202	3.96850	8.54988	.160000
26	39.1876	245.314	2.50200	7.91202	1.84300	3.97062	8.55444	.159744
27	39.3129	246.492	2.50400	7.91833	1.84398	3.97273	8.55899	.159490
28	39.4384	247.673	2.50599	7.92465	1.84496	3.97484	8.56354	.159236
6.29	39.5641	248.858	2.50799	7.93095	1.84594	3.97695	8.56808	.158983
6.30	39.6900	250.047	2.50998	7.93725	1.84691	3.97906	8.57262	.158730
31	39.8161	251.240	2.51197	7.94355	1.84789	3.98116	8.57715	.158479
32	33.9424	252.436	2.51396	7.94984	1.84887	3.98326	8.58168	.158228
33	40.0689	253.636	2.51595	7.95613	1.84984	3.98536	8.58620	.157978
6.34	40.1956	254.840	2.51794	7.96241	1.85082	3.98746	8.59072	.157729
6.35	40.3225	256.048	2.51992	7.96869	1.85179	3.98956	8.59524	.157480
36	40.4496	257.259	2.52190	7.97496	1.85276	3.99165	8.59975	.157233
37	40.5769	258.475	2.52389	7.98123	1.85373	3.99374	8.60425	.156986
38	40.7044	259.694	2.52587	7.98749	1.85470	3.99583	8.60875	.156740
6.39	40.8321	260.917	2.52784	7.99375	1.85567	3.99792	8.61325	.156495
6.40	40.9600	262.144	2.52982	8.00000	1.85664	4.00000	8.61774	.156250
41	41.0881	263.375	2.53180	8.00625	1.85760	4.00208	8.62222	.156006
.42	41.2164	264.609	2.53377	8.01249	1.85857	4.00416	8.62671	.155763
.43	41.3449	265.848	2.53574	8.01873	1.85953	4.00624	8.63118	.155521
6.44	41.4736	267.090	2.53772	8.02496	1.86050	4.00832	8.63566	.155280
6.45	41.6025	268.336	2.53969	8.03119	1.86146	4.01039	8.64012	.155039
.	41.7316	269.586	2.54165	8.03741	1.86242	4.01246	8.64459	.154799
	41 8609	270.840	2.54362	8.04363	1.86338	4.01453	8.64904	.154560
	41.9904	272.098	2.54558	8.04984	1.86434	4.01660	8.65350	.154321
6.46	42.1201	273.359	2.54755	8.05605	1.86530	4.01866	8.65795	.154083
6.48	42.2500	274.625	2.54951	8.06226	1.86626	4.02073	8.66239	.153846

n	n^2	n^3	$\sqrt{n}$	$\sqrt{10\,n}$	$\sqrt[3]{n}$	$\sqrt[3]{10\,n}$	$\sqrt[3]{100\,n}$	$\dfrac{1}{n}$
6.51	42 3801	275.894	2.55147	8.06846	1.86721	4.02279	8.66683	.153610
6 52	42.5104	277.168	2.55343	8.07465	1.86817	4.02485	8.67127	.153374
6 53	42 6409	278.445	2.55539	8.08084	1.86912	4.02690	8.67570	.153139
6.54	42.7716	279.726	2.55734	8.08703	1.87008	4.02896	8.68012	.152905
6.5	42.9025	281.011	2.55930	8.09321	1.87103	4.03101	8.68455	.152672
6.56	43.0336	282.300	2.56125	8.09938	1.87198	4.03306	•8.68896	.152439
6 57	43.1649	283.593	2.56320	8.10555	1.87293	4.03511	8.69338	.152207
6 58	43.2964	284 890	2.56515	8.11172	1.87388	4.03715	8.69778	.151976
6 59	43.4281	286.191	2.56710	8.11788	1.87483	4.03920	8.70219	.151745
6.60	43.5600	287.496	2.56905	8.12404	1.87578	4.04124	8.70659	.151515
1	43.6921	288.805	2.57099	8.13019	1.87672	4.04328	8.71098	.151286
62	43.8244	290.118	2.57294	8.13634	1.87767	4.04532	8.71537	.151057
63	43 9569	291.434	2.57488	8.14248	1.87862	4.04735	8.71976	.150830
6 64	44 0896	292.755	2.57682	8.14862	1.87956	4.04939	8.72414	.150602
6.65	44.2225	294.080	2.57876	8.15475	1.88050	4.05142	8.72852	.150376
66	44.3556	295.408	2.58070	8.16088	1.88144	4.05345	8.73289	.150150
67	44.4889	296.741	2.58263	8.16701	1.88239	4.05548	8.73726	.149925
68	44.6224	298.078	2.58457	8.17313	1.88333	4.05750	8.74162	.149701
6 69	44.7561	299.418	2.58650	8.17924	1.88427	4.05953	8.74598	.149477
6.70	44.8900	300.763	2.58844	8.18535	1.88520	4.06155	8.75034	.149254
.71	45.0241	302.112	2.59037	8.19146	1.88614	4.06357	8.75469	.149031
.72	45 1584	303.464	2.59230	8.19756	1.88708	4.06558	8.75904	.148810
.73	45 2929	304.821	2.59422	8.20366	1.88801	4.06760	8.76338	.148588
6.74	45.4276	306.182	2.59615	8.20975	1.88895	4.06961	8.76772	.148368
6.75	45.5625	307.547	2.59808	8.21584	1.88988	4.07163	8.77205	.148148
.	45.6976	308.916	2.60000	8.22192	1.89081	4.07364	8.77638	.147929
.	45.8329	310 289	2.60192	8.22800	1.89175	4.07564	8.78071	.147711
.	45 9684	311.666	2.60384	8.23408	1.69268	4.07765	8.78503	.147493
6.76	46.1041	313.047	2.60576	8.24015	1.89361	4.07965	8.78935	.147275
6.80	46.2400	314.432	2.60768	8.24621	1.89454	4.08166	8.79366	.147059
.	46.3761	315.821	2.60960	8.25227	1.89546	4.08365	8.79797	.146843
.	46.5124	317.215	2.61151	8.25833	1.89639	4.08565	8.80227	.146628
.	46.6489	318 612	2.61343	8.26438	1.89732	4.08765	8.80657	.146413
6.81	46 7856	320.014	2.61534	8.27043	1.89824	4.08964	8.81087	.146199
6.84	46.9225	321.419	2.61725	8.27647	1.89917	4.09164	8.81516	.145985
.	47 0596	322.829	2.61916	8.28251	1.90009	4.09362	8.81945	.145773
.	47.1969	324 243	2.62107	8.28855	1.90102	4.09561	8.82373	.145560
.	47.3344	325.661	2.62298	8.29458	1.90194	4.09760	8.82801	.145349
6.86	47.4721	327.083	2.62488	8.30060	1.90286	4.09958	8.83229	.145138
6.80	47.6100	328.509	2.62679	8.30662	1.90378	4.10157	8.83656	.144928
.	47.7481	329.939	2.62869	8.31264	1.90470	4.10355	8.84082	.144718
.	47.8864	331.374	2.63059	8.31865	1.90562	4.10552	8.84509	.144509
.	48 0249	332.813	2.63249	8.32466	1.90653	4.10750	8.84934	.144300
6.91	48.1636	334.255	2.63439	8.33067	1.90745	4.10948	8.85360	.144092
6.98	48.3025	335.702	2.63629	8.33667	1.90837	4.11145	8.85785	.143885
.	48.4416	337.154	2.63818	8.34266	1.90928	4.11342	8.86210	.143678
.	48.5809	338.609	2.64008	8.34865	1.91019	4.11539	8.86634	.143472
.	48.7204	340.068	2.64197	8.35464	1.91111	4.11736	8.87058	.143267
6.96	48.8601	341.532	2.64386	8.36062	1.91202	4.11932	8.87481	.143062
6.00	49.0000	343.000	2.64575	8.36660	1.91293	4.12129	8.87904	.142857

n	n^2	n^3	$\sqrt{n}$	$\sqrt{10\,n}$	$\sqrt[3]{n}$	$\sqrt[3]{10\,n}$	$\sqrt[3]{100\,n}$	$\dfrac{1}{n}$
7.01	49.1401	344.472	2.64764	8.37257	1.91384	4.12325	8.88327	.142653
7	49.2804	345.948	2.64953	8.37854	1.91475	4.12521	8.88749	.142450
7	49.4209	347.429	2.65141	8.38451	1.91566	4.12716	8.89171	.142248
7.02	49.5616	348.914	2.65330	8.39047	1.91657	4.12912	8.89592	.142046
7.03	49.7025	350.403	2.65518	8.39643	1.91747	4.13107	8 90013	.141844
7	49.8436	351.896	2.65707	8.40238	1.91838	4.13303	8 90434	.141643
7	49.9849	353.393	2.65895	8.40833	1.91929	4 13498	8.90854	.141443
7.06	50.1264	354.895	2.66083	8.41427	1.92019	4 13695	8.91274	.141243
7.08	50.2681	356.401	2.66271	8.42021	1.92109	4.13887	8.91693	.141044
7.10	50.4100	357.911	2.66458	8.42615	1.92200	4.14082	8 92112	.140845
7.11	50.5521	359.425	2.66646	8.43208	1.92290	4.14276	8 92531	.140647
7.12	50.6944	360.944	2.66833	8.43801	1.92380	4 14470	8 92949	.140449
7.13	50 8369	362.467	2.67021	8.44393	1.92470	4.14664	8 93367	.140253
7.14	50.9796	363.994	2.67208	8.44985	1.92560	4.14858	8.93784	.140056
7.15	51.1225	365.526	2.67395	8.45577	1.92650	4.15051	8.94201	.139860
7.16	51.2656	367.062	2.67582	8.46168	1.92740	4.15245	8.94618	.139665
7.17	51.4089	368.602	2.67769	8.46759	1.92829	4.15438	8 95034	.139470
7.18	51.5524	370.146	2.67955	8.47349	1.92919	4.15631	8.95450	.139276
7.19	51.6961	371.695	2.68142	8.47939	1.93008	4.15824	8.95866	.139082
7.20	51.8400	373.248	2.68328	8.48528	1.93098	4.16017	8.96281	.138889
7.21	51.9841	374.805	2.68514	8.49117	1.93187	4.16209	8 96696	.138696
7 22	52.1284	376.367	2.68701	8.49706	1.93277	4.16402	8.97110	.138504
7 23	52.2729	377.933	2.68887	8.50294	1.93366	4.16594	8.97524	.138313
7.24	52.4176	379.503	2 69072	8.50882	1.93455	4.16786	8.97938	.138122
7.25	52.5625	381.078	2.69258	8.51469	1.93544	4.16978	8.98351	.137931
7.26	52.7076	382.657	2.69444	8.52056	1.93633	4.17169	8 98764	.137741
7.27	52.8529	384.241	2.69629	8.52643	1.93722	4.17361	8.99176	.137552
7.28	52.9984	385.828	2.69815	8.53229	1.93810	4.17552	8.99588	.137363
7.29	53.1441	387.420	2.70000	8.53815	1.93899	4.17743	9.00000	.137174
7.30	53.2900	389.017	2.70185	8.54400	1.93988	4.17934	9.00411	.136986
7.	53.4361	390.618	2.70370	8.54985	1.94076	4.18125	9.00822	.136799
7.	53.5824	392.223	2.70555	8.55570	1.94165	4.18315	9.01233	.136612
7.31	53.7289	393.833	2.70740	8.56154	1.94253	4.18506	9.01643	.136426
7.34	53.8756	395.447	2.70924	8.56738	1.94341	4.18696	9.02053	.136240
7.35	54.0225	397.065	2.71109	8.57321	1.94430	4.18886	9.02462	.136054
7.	54.1696	398.688	2.71293	8.57904	1.94518	4.19076	9.02871	.135870
7.	54.3169	400.316	2.71477	8.58487	1.94606	4.19266	9 03280	.135685
7.	54.4644	401.947	2.71662	8 59069	1.94694	4.19455	9 03689	.135501
7.36	54.6121	403.583	2.71846	8.59651	1.94782	4.19644	9.04097	.135318
7.08	54.7600	405.224	2.72029	8.60233	1.94870	4.19834	9.04504	.135135
7.41	54.9081	406.869	2.72213	8.60814	1.94957	4.20023	9.04911	.134953
7.42	55.0564	408.518	2 72397	8.61394	1.95045	4 20212	9.05318	.134771
7.43	55.2049	410.172	2.72580	8.61974	1.95132	4 20400	9.05725	.134590
7.44	55.3536	411.831	2.72764	8.62554	1.95220	4.20589	9.06131	.134409
7.45	55.5025	413.494	2.72947	8.63134	1.95307	4.20777	9.06537	.134228
7.	55.6516	415.161	2.73130	8 63713	1.95395	4.20965	9.06942	.134048
7.	55.8009	416.833	2.73313	8.64292	1.95482	4.21153	9 07347	.133869
7.	55.9504	418.509	2.73496	8 64870	1.95569	4.21341	9.07752	.133690
7.46	56.1001	420.190	2.73679	8.65448	1 95656	4.21529	9.08156	.133511
7.58	56.2500	421.875	2.73861	8.66025	1.95743	4.21716	9.08560	.133333

n	n^2	n^3	$\sqrt{n}$	$\sqrt{10\,n}$	$\sqrt[3]{n}$	$\sqrt[3]{10\,n}$	$\sqrt[3]{100\,n}$	$\dfrac{1}{n}$
7.51	56.4001	423.565	2.74044	8.66603	1.95830	4.21904	9.08964	.133156
7.52	56.5504	425.259	2.74226	8.67179	1.95917	4.22091	9.09367	.132979
7.53	56.7009	426.958	2.74408	8.67756	1.96004	4.22278	9.09770	.132802
7.54	56.8516	428.661	2.74591	8.68332	1.96091	4.22465	9.10173	.132626
7.55	57.0025	430.369	2.74773	8.68907	1.96177	4.22651	9.10575	.132450
7	57.1536	432.081	2.74955	8.69483	1.96264	4.22838	9.10977	.132275
7	57.3049	433.798	2.75136	8.70057	1.96350	4.23024	9.11378	.132100
7	57.4564	435.520	2.75318	8.70632	1.96437	4.23210	9.11779	.131926
7.56	57.6081	437.245	2.75500	8.71206	1.96523	4.23396	9.12180	.131752
7:56	57.7600	438.976	2.75681	8.71780	1.96610	4.23582	9.12581	.131579
7	57.9121	440.711	2.75862	8.72353	1.96696	4.23768	9.12981	.131406
7	58.0644	442.451	2.76043	8.72926	1.96782	4.23954	9.13380	.131234
7	58.2169	444.195	2.76225	8.73499	1.96868	4.24139	9.13780	.131062
7.61	58.3696	445.994	2.76405	8.74071	1.96954	4.24324	9.14179	.130890
7.62	58.5225	447.697	2.76586	8.74643	1.97040	4.24509	9.14577	.130719
7	58.6756	449.455	2.76767	8.75214	1.97126	4.24694	9.14976	.130548
7.	58.8289	451.218	2.76948	8.75785	1.97211	4.24879	9.15374	.130378
7	58.9824	452.985	2.77128	8.76356	1.97297	4.25063	9.15771	.130208
7.66	59.1361	454.757	2.77308	8.76926	1.97383	4.25248	9.16169	.130039
7.66	59.2900	456.533	2.77489	8.77496	1.97468	4.25432	9.16566	.129870
7.71	59.4441	458.314	2.77669	8.78066	1.97554	4.25616	9.16962	.129702
7.72	59.5984	460.100	2.77849	8.78635	1.97639	4.25800	9.17359	.129534
.73	59.7529	461.890	2.78029	8.79204	1.97724	4.25984	9.17754	.129366
.74	59.9076	463.685	2.78209	8.79773	1.97809	4.26168	9.18150	.129199
7.75	60.0625	465.484	2.78388	8.80341	1.97895	4.26351	9.18545	.129032
7.76	60.2176	467.289	2.78568	8.80909	1.97980	4.26534	9.18940	.128866
7.77	60.3729	469.097	2.78747	8.81476	1.98065	4.26717	9.19335	.128700
7.78	60.5284	470.911	2.78927	8.82043	1.98150	4.26900	9.19729	.128535
7.79	60.6841	472.729	2.79106	8.82610	1.98234	4.27083	9.20123	.128370
7.80	60.8400	474.552	2.79285	8.83176	1.98319	4.27266	9.20516	.128205
7	60.9961	476.380	2.79464	8.83742	1.98404	4.27448	9.20910	.128041
7	61.1524	478.212	2.79643	8.84308	1.98489	4.27631	9.21303	.127877
7	61.3089	480 049	2.79821	8.84873	1.98573	4.27813	9.21695	.127714
7.81	61.4656	481 890	2.80000	8.85438	1.98658	4.27995	9.22087	.127551
7.85	61.6225	483.737	2.80179	8.86002	1.98742	4.28177	9.22479	.127389
7.86	61.7796	485.588	2.80357	8.86566	1.98826	4.28359	9.22871	.127227
7.87	61.9369	487.443	2.80535	8.87130	1.98911	4.28540	9.23262	.127065
7.88	62.0944	489.304	2.80713	8.87694	1.98995	4.28722	9.23653	.126904
7.89	62.2521	491.169	2.80891	8.88257	1.99079	4.28903	9.24043	.126743
7.90	62.4100	493.039	2.81069	8.88819	1.99163	4.29084	9.24433	.126582
7.91	62.5681	494.914	2.81247	8.89382	1.99247	4.29265	9.24823	.126422
7.92	62.7264	496.793	2.81425	8.89944	1.99331	4.29446	9.25213	.126263
7.93	62.8849	498.677	2.81603	8.90505	1.99415	4.29627	9.25602	.126103
7.94	63.0436	500.566	2.81780	8.91067	1.99499	4.29807	9.25991	.125945
7.95	63.2025	502.460	2.81957	8.91628	1.99582	4.29987	9.26380	.125786
	63.3616	504.358	2.82135	8.92188	1.99666	4.30168	9.26768	.125628
	63.5209	506.262	2.82312	8.92749	1.99750	4.30348	9.27156	.125471
	63.6804	508.170	2.82489	8.93308	1.99833	4.30528	9.27544	.125313
7.96	63.8401	510.082	2.82666	8.93868	1.99917	4.30707	9.27931	.125156
8.00	64.0000	512.000	2.82843	8.94427	2.00000	4.30887	9.28318	.125000

n	n^2	n^3	$\sqrt{n}$	$\sqrt{10\,n}$	$\sqrt[3]{n}$	$\sqrt[3]{10\,n}$	$\sqrt[3]{100\,n}$	$\dfrac{1}{n}$
01	64.1601	513.922	2.83019	8.94986	2.00083	4.31066	9.28704	.124844
	64.3204	515.850	2.83196	8.95545	2.00167	4.31246	9.29091	.124688
	64.4809	517.782	2.83373	8.96103	2.00250	4.31425	9.29477	.124533
02	64.6416	519.718	2.83549	8.96660	2.00333	4.31604	9.29862	.124378
8.05	64.8025	521.660	2.83725	8.97218	2.00416	4.31783	9.30248	.124224
	64.9636	523.607	2.83901	8.97775	2.00499	4.31961	9.30633	.124070
	65.1249	525.558	2.84077	8.98332	2.00582	4.32140	9.31018	.123916
	65.2864	527.514	2.84253	8.98888	2.00664	4.32318	9.31402	.123762
06	65.4481	529.475	2.84429	8.99444	2.00747	4.32497	9.31786	.123609
8.00	65.6100	531.441	2.84605	9.00000	2.00830	4.32675	9.32170	.123457
8.11	65.7721	533.412	2.84781	9.00555	2.00912	4.32853	9.32553	.123305
.12	65.9344	535.387	2.84956	9.01110	2.00995	4.53031	9.32936	.123153
8.13	66.0969	537.368	2.85132	9.01665	2.01078	4.33208	9.33319	.123001
8.14	66.2596	539.353	2.85307	9.02219	2.01160	4.33386	9.33702	.122850
8.15	66.4225	541.343	2.85482	9.02774	2.01242	4.33563	9.34084	.122699
8.	66.5856	543.338	2.85657	9.03327	2.01325	4.33741	9.34466	.122549
8.	66.7489	545.339	2.85832	9.03881	2.01407	4.33918	9.34847	.122399
8.	66.9124	547.343	2.86007	9.04434	2.01489	4.34095	9.35229	.122249
8.	67.0761	549.353	2.86182	9.04986	2.01571	4.34272	9.35610	.122100
8.20	67.2400	551.368	2.86356	9.05539	2.01653	4.34448	9.35990	.121951
8.	67.4041	553.388	2.86531	9.06091	2.01735	4.34625	9.36370	.121803
8.	67.5684	555.412	2.86705	9.06642	2.01817	4.34801	9.36751	.121655
8.	67.7329	557.442	2.86880	9.07193	2.01899	4.34977	9.37130	.121507
.21	67.8976	559.476	2.87054	9.07744	2.01980	4.35153	9.37510	.121359
8.25	68.0625	561.516	2.87228	9.08295	2.02062	4.35329	9.37889	.121212
.26	68.2276	563.560	2.87402	9.08845	2.02144	4.35505	9.38268	.121065
.27	68.3929	565.609	2.87576	9.09395	2.02225	4.35681	9.38646	.120919
.28	68.5584	567.664	2.87750	9.09945	2.02307	4.35856	9.39024	.120773
.29	68.7241	569.723	2.87924	9.10494	2.02388	4.36032	9.39402	.120627
8.30	68.8900	571.787	2.88097	9.11043	2.02469	4.36207	9.39780	.120482
.	69.0561	573.856	2.88271	9.11592	2.02551	4.36382	9.40157	.120337
.	69.2224	575.930	2.88444	9.12140	2.02632	4.36557	9.40534	.120192
.	69.3889	578.010	2.88617	9.12688	2.02713	4.36732	9.40911	.120048
.	69.5556	580.094	2.88791	9.13236	2.02794	4.36907	9.41287	.119904
8.35	69.7225	582.183	2.88964	9.13783	2.02875	4.37081	9.41663	.119761
.	69.8896	584.277	2.89137	9.14330	2.02956	4.37255	9.42039	.119617
	70.0569	586.376	2.89310	9.14877	2.03037	4.37430	9.42414	.119474
	70.2244	588.480	2.89482	9.15423	2.03118	4.37604	9.42789	.119332
36	70.3921	590.590	2.89655	9.15969	2.03199	4.37778	9.43164	.119190
8.40	70.5600	592.704	2.89828	9.16515	2.03279	4.37952	9.43539	.119048
	70.7281	594.823	2.90000	9.17061	2.03360	4.38126	9.43913	.118906
	70.8964	596.948	2.90172	9.17606	2.03440	4.38299	9.44287	.118765
	71.0649	599.077	2.90345	9.18150	2.03521	4.38473	9.44661	.118624
.41	71.2336	601.212	2.90517	9.18695	2.03601	4.38646	9.45034	.118483
8.45	71.4025	603.351	2.90689	9.19239	2.03682	4.38819	9.45407	.118343
	71.5716	605.496	2.90861	9.19783	2.03762	4.38992	9.45780	.118203
	71.7409	607.645	2.91033	9.20326	2.03842	4.39165	9.46152	.118064
	71.9104	609.800	2.91204	9.20869	2.03923	4.39338	9.46525	.117925
46	72.0801	611.960	2.91376	9.21412	2.04003	4.39511	9.46897	.117786
8.50	72.2500	614.125	2.91548	9.21954	2.04083	4.39683	9.47268	.117647

n	n^2	n^3	$\sqrt{n}$	$\sqrt{10\,n}$	$\sqrt[3]{n}$	$\sqrt[3]{10\,n}$	$\sqrt[3]{100\,n}$	$\dfrac{1}{n}$
	72.4201	616.295	2.91719	9.22497	2.04163	4.39855	9.47640	.117509
	72.5904	618.470	2.91890	9.23038	2.04243	4.40028	9.48011	.117371
	72.7609	620.650	2.92062	9.23580	2.04323	4.40200	9.48381	.117233
	72.9316	622.836	2.92233	9.24121	2.04402	4.40372	9.48752	.117096
8.55	73.1025	625.026	2.92404	9.24662	2.04482	4.40543	9.49122	.116959
56	73.2736	627.222	2 92575	9.25203	2.04562	4.40715	9.49492	.116822
57	73.4449	629.423	2.92746	9.25743	2.04641	4.40887	9.49861	.116686
58	73.6164	631.629	2.92916	9.26283	2.04721	4.41058	9.50231	.116550
.59	73.7881	633.840	2.93087	9.26823	2.04801	4.41229	9.50600	.116414
8.60	73.9600	636.056	2.93258	9.27362	2.04880	4.41400	9.50969	.116279
8.	74.1321	638.277	2.93428	9.27901	2.04959	4.41571	9.51337	.116144
8.	74.3044	640.504	2.93598	9.28440	2.05039	4.41742	9.51705	.116009
8.61	74.4769	642.736	2.93769	9.28978	2.05118	4.41913	9.52073	.115875
8.64	74.6496	644.973	2.93939	9.29516	2.05197	4.42084	9.52441	.115741
8.65	74.8225	647.215	2.94109	9.30054	2.05276	4.42254	9.52808	.115607
8.	74.9956	649.462	2.94279	9.30591	2.05355	4.42425	9.53175	.115473
8.	75.1689	651.714	2.94449	9.31128	2.05434	4.42595	9.53542	.115340
8.	75.3424	653.972	2 94618	9.31665	2.05513	4.42765	9.53908	.115207
8.66	75.5161	656.235	2.94788	9.32202	2.05592	4.42935	9.54274	.115075
8.68	75.6900	658.503	2.94958	9.32738	2.05671	4.43105	9.54640	.114943
8.71	75.8641	660.776	2.95127	9.33274	2.05750	4.43274	9.55006	.114811
8.72	76.0384	663.055	2.95296	9.33809	2.05828	4.43444	9.55371	.114679
8.73	76.2129	665.339	2.95466	9.34345	2.05907	4.43614	9.55736	.114548
8.74	76.3876	667.628	2.95635	9.34880	2.05986	4.43783	9.56101	.114417
8.75	76.5625	669.922	2.95804	9.35414	2.06064	4.43952	9.56466	.114286
8.	76.7376	672.221	2.95973	9.35949	2.06143	4.44121	9.56830	.114155
8.	76.9129	674.526	2.96142	9.36483	2.06221	4.44290	9.57194	.114025
8.	77.0884	676.836	2.96311	9.37017	2 06299	4.44459	9.57557	.113895
8.76	77.2641	679.151	2.96479	9.37550	2.06378	4.44627	9.57921	.113766
8.80	77.4400	681.472	2.96648	9.38083	2.06456	4.44796	9.58284	.113636
8.	77.6161	683.798	2.96816	9.38616	2.06534	4.44964	9.58647	.113507
8.	77.7924	686.129	2.96985	9.39149	2.06612	4.45133	9.59009	.113379
8.	77.9689	688.465	2.97153	9.39681	2.06690	4.45301	9.59372	.113250
8.81	78.1456	690.807	2.97321	9.40213	2.06768	4.45469	9.59734	.113122
8.88	78.3225	693.154	2.97489	9.40744	2.06846	4.45637	9.60095	.112994
8.	78.4996	695.506	2.97658	9.41276	2.06924	4.45805	9.60457	.112867
8.	78.6769	697.864	2.97825	9.41807	2.07002	4.45972	9.60818	.112740
8.	78.8544	700.227	2.97993	9.42338	2.07080	4.46140	9.61179	.112613
8.86	79.0321	702.595	2.98161	9.42868	2.07157	4.46307	9.61540	.112486
8.88	79.2100	704.969	2.98329	9.43398	2.07235	4.46474	9.61900	.112360
8.91	79.3881	707.348	2.98496	9.43928	2.07313	4.46642	9.62260	.112233
	79.5664	709.732	2.98664	9.44458	2.07390	4.46809	9 62620	.112108
8.	79.7449	712.122	2.98831	9.44987	2.07468	4.46976	9.62980	.111982
8.92	79.9236	714.517	2.98998	9.45516	2.07545	4.47142	9.63339	.111857
8.93	80.1025	716.917	2.99166	9.46044	2.07622	4.47309	9.63698	.111732
8.	80.2816	719.323	2.99333	9.46573	2.07700	4.47476	9.64057	.111607
8.	80.4609	721.734	2.99500	9.47101	2.07777	4.47642	9.64415	.111483
8.96	80.6404	724.151	2.99666	9.47629	2.07854	4.47808	9.64774	.111359
8.98	80.8201	726.573	2.99833	9.48156	2.07931	4.47974	9.65132	.111235
9.00	81.0000	729.000	3.00000	9.48683	2.08008	4.48140	9.65489	.111111

n	n^2	n^3	$\sqrt{n}$	$\sqrt{10\,n}$	$\sqrt[3]{n}$	$\sqrt[3]{10\,n}$	$\sqrt[3]{100\,n}$	$\dfrac{1}{n}$
9.01	81.1801	731.433	3.00167	9.49210	2.08085	4.48306	9.65847	.110988
9.02	81.3604	733.871	3.00333	9.49737	2.08162	4.48472	9 66204	.110865
9.03	81.5409	736.314	3.00500	9.50263	2.08239	4.48638	9.66561	.110742
9.04	81.7216	738.763	3.00666	9.50789	2.08316	4.48803	9.66918	.110620
9.05	81.9025	741.218	3.00832	9.51315	2.08393	4.48968	9.67274	.110497
9.06	82.0836	743.677	3.00998	9.51840	2.08470	4.49134	9.67630	.110375
9.07	82.2649	746.143	3.01164	9.52365	2.08546	4.49299	9.67986	.110254
9.08	82.4464	748.613	3.01330	9.52890	2.08623	4.49464	9.68342	.110132
9.09	82.6281	751.089	3.01496	9.53415	2.08699	4.49629	9.68697	.110011
9.10	82.8100	753.571	3.01662	9.53939	2.08776	4.49794	9.69052	.109890
9.11	82.9921	756.058	3.01828	9.54463	2.08852	4.49959	9.69407	.109770
9.12	83.1744	758.551	3.01993	9.54987	2.08929	4.50123	9.69762	.109649
9.13	83.3569	761.048	3.02159	9.55510	2.09005	4.50288	9.70116	.109529
9.14	83.5396	763.552	3.02324	9.56033	2.09081	4.50452	9.70470	.109409
9.15	83.7225	766.061	3.02490	9.56556	2.09158	4.50616	9.70824	.109290
9.16	83.9056	768.575	3.02655	9.57079	2.09234	4.50780	9.71177	.109170
9.17	84.0889	771.095	3.02820	9.57601	2.09310	4.50945	9.71531	.109051
9.18	84.2724	773.621	3.02985	9.58123	2.09386	4.51108	9.71884	.108933
9.19	84.4561	776.152	3.03150	9.58645	2.09462	4.51272	9.72236	.108814
9.20	84.6400	778.688	3.03315	9.59166	2.09538	4.51436	9.72589	.108696
9.21	84.8241	781.230	3.03480	9.59687	2.09614	4.51599	9.72941	.108578
9.22	85.0084	783.777	3.03645	9.60208	2.09690	4.51763	9.73293	.108460
9.23	85.1929	786.330	3.03809	9.60729	2.09765	4.51926	9.73645	.108342
9.24	85.3776	788.889	3.03974	9.61249	2.09841	4.52089	9.73996	.108225
9.25	85.5625	791.453	3.04138	9.61769	2.09917	4.52252	9.74348	.108108
9.2	85.7476	794.023	3.04302	9.62289	2.09992	4.52415	9.74699	.107991
9.2	85.9329	796.598	3.04467	9.62808	2.10068	4.52578	9.75049	.107875
9.28	86.1184	799.179	3.04631	9.63328	2.10144	4.52740	9.75400	.107759
9.29	86.3041	801.765	3.04795	9.63846	2.10219	4.52903	9.75750	.107643
9.30	86.4900	804.357	3.04959	9.64365	2.10294	4.53065	9.76100	.107527
9.	86.6761	806.954	3.05123	9.64883	2.10370	4.53228	9.76450	.107411
9.	86.8624	809.558	3.05287	9.65401	2.10445	4.53390	9.76799	.107296
9.	87.0489	812.166	3.05450	9.65919	2.10520	4.53552	9.77148	.107181
9.31	87.2356	814.781	3.05614	9.66437	2.10595	4.53714	9.77497	.107066
9.33	87.4225	817.400	3.05778	9.66954	2.10671	4.53876	9.77846	.106952
9.	87.6096	820.026	3.05941	9.67471	2.10746	4.54038	9.78195	.106838
9.	87.7969	822.657	3.06105	9.67988	2.10821	4.54199	9.78543	.106724
9.	87.9844	825.294	3.06268	9.68504	2.10896	4.54361	9.78891	.106610
9.36	88.1721	827.936	3.06431	9.69020	2.10971	4.54522	9.79239	.106496
9.88	88.3600	830.584	3.06594	9.69536	2.11045	4.54684	9.79586	.106383
41	88.5481	833.238	3.06757	9.70052	2.11120	4.54845	9.79933	.106270
42	88.7364	835.897	3.06920	9.70567	2.11195	4.55006	9.80280	.106157
43	88.9249	838.562	3.07083	9.71082	2.11270	4.55167	9.80627	.106045
44	89.1136	841.232	3.07246	9.71597	2.11344	4.55328	9.80974	.105932
9.45	89.3025	843.909	3.07409	9.72111	2.11419	4.55488	9.81320	.105820
	89.4916	846.591	3.07571	9.72625	2.11494	4.55649	9.81666	.105708
	89.6809	849.278	3.07734	9.73139	2.11568	4.55809	9.82012	.105597
	89.8704	851.971	3.07896	9.73653	2.11642	4.55970	9.82357	.105485
46	90.0601	854.670	3.08058	9.74166	2.11717	4.56130	9.82703	.105374
9.60	90.2500	857.375	3.08221	9.74679	2.11791	4.56290	9.83048	.105263

n	n^2	n^3	$\sqrt{n}$	$\sqrt{10\,n}$	$\sqrt[3]{n}$	$\sqrt[3]{10\,n}$	$\sqrt[3]{100\,n}$	$\dfrac{1}{n}$
9.51	90.4401	860.085	3.08383	9.75192	2.11865	4.56450	9.83392	.105153
9.52	90.6304	862.801	3.08545	9.75705	2.11940	4.56610	9.83737	.105042
9.53	90.8209	865.523	3.08707	9.76217	2.12014	4.56770	9.84081	.104932
9.54	91.0116	868.251	3.08869	9.76729	2.12088	4.56930	9.84425	.104822
9.55	91.2025	870.984	3.09031	9.77241	2.12162	4.57089	9.84769	.104712
9.56	91.3936	873.723	3.09192	9.77753	2.12236	4.57249	9.85113	.104603
9.57	91.5849	876.467	3.09354	9.78264	2.12310	4.57408	9.85456	.104493
9.58	91.7764	879.218	3.09516	9.78775	2.12384	4.57568	9.85799	.104384
9.59	91.9681	881.974	3.09677	9.79285	2.12458	4.57727	9.86142	.104275
9.60	92.1600	884.736	3.09839	9.79796	2.12582	4.57886	9.86485	.104167
9.61	92.3521	887.504	3.10000	9.80306	2.12605	4.58045	9.86827	.104058
9.62	92.5444	890.277	3.10161	9.80816	2.12679	4.58203	9.87169	.103950
9.63	92.7369	893.056	3.10322	9.81326	2.12753	4.58362	9.87511	.103842
9.64	92.9296	895.841	3.10483	9.81835	2.12826	4.58521	9.87853	.103734
9.65	93.1225	898.632	3.10644	9.82344	2.12900	4.58679	9.88195	.103627
9.66	93.3156	901.429	3.10805	9.82853	2.12974	4.58838	9.88536	.103520
9.67	93.5089	904.231	3.10966	9.83362	2.13047	4.58996	9.88877	.103413
9.68	93.7024	907.039	3.11127	9.83870	2.13120	4.59154	9.89217	.103306
9.69	93.8961	909.853	3.11288	9.84378	2.13194	4.59312	9.89558	.103199
9.70	94.0900	912.673	3.11448	9.84886	2.13267	4.59470	9.89898	.103093
9.71	94.2841	915.499	3.11609	9.85393	2.13340	4.59628	9.90238	.102987
9.72	94.4784	918.330	3.11769	9.85901	2.13414	4.59786	9.90578	.102881
9.73	94.6729	921.167	3.11929	9.86408	2.13487	4.59943	9.90918	.102775
9.74	94.8676	924.010	3.12090	9.86914	2.13560	4.60101	9.91257	.102669
9.75	95.0625	926.859	3.12250	9.87421	2.13633	4.60258	9.91596	.102564
9.76	95.2576	929.714	3.12410	9.87927	2.13706	4.60416	9.91935	.102459
9.77	95.4529	932.515	3.12570	9.88433	2.13779	4.60573	9.92274	.102354
9.78	95.6484	935.441	3.12730	9.88939	2.13852	4.60730	9.92612	.102250
9.79	95.8441	938.314	3.12890	9.89444	2.13925	4.60887	9.92950	.102145
9.80	96.0400	941.192	3.13050	9.89949	2.13997	4.61044	9.93288	.102041
9.81	96.2361	944.076	3.13209	9.90454	2.14070	4.61200	9.93626	.101937
9.82	96.4324	946.966	3.13369	9.90959	2.14143	4.61357	9.93964	.101833
9.83	96.6289	949.862	3.13528	9.91464	2.14216	4.61513	9.94301	.101729
9.84	96.8256	952.764	3.13688	9.91968	2.14288	4.61670	9.94638	.101626
9.85	97.0225	955.672	3.13847	9.92472	2.14361	4.61826	9.94975	.101523
9.86	97.2196	958.585	3.14006	9.92975	2.14433	4.61983	9.95311	.101420
9.87	97.4169	961.505	3.14166	9.93479	2.14506	4.62139	9.95648	.101317
9.88	97.6144	964.430	3.14325	9.93982	2.14578	4.62295	9.95984	.101215
9.89	97.8121	967.362	3.14484	9.94485	2.14651	4.62451	9.96320	.101112
9.90	98.0100	970.299	3.14643	9.94987	2.14723	4.62607	9.96655	.101010
9.91	98.2081	973.242	3.14802	9.95490	2.14795	4.62762	9.96991	.100908
9.92	98.4064	976.191	3.14960	9.95992	2.14867	4.62918	9.97326	.100807
9.93	98.6049	979.147	3.15119	9.96494	2.14940	4.63073	9.97661	.100705
9.94	98.8036	982.108	3.15278	9.96995	2.15012	4.63229	9.97996	.100604
9.95	99.0025	985.075	3.15436	9.97497	2.15084	4.63384	9.98331	.100503
9.96	99.2016	988.048	3.15595	9.97998	2.15156	4.63539	9.98665	.100402
9.97	99.4009	991.027	3.15753	9.98499	2.15228	4.63694	9.98999	.100301
9.98	99.6004	994.012	3.15911	9.98999	2.15300	4.63849	9.99333	.100200
9.99	99.8001	997.003	3.16070	9.99500	2.15372	4.64004	9.99667	.100100
10.00	100.000	1000.00	3.16228	10.0000	2.15443	4.64159	10.0000	.100000

FORMULAS

A *formula* is a brief statement of a rule, in which letters or other symbols are used to denote the different quantities involved. For example, the rule for finding the volume of a rectangular prism is as follows: *The volume of a rectangular prism is equal to the product of the length, width, and height of the prism.* If the dimensions of the prism are taken in inches, the volume will be in cubic inches; if they are taken in feet, the volume will be in cubic feet; and so on. Suppose, however, that the volume is denoted by v, the length by l, the width by w, and the height by h. Then, the foregoing rule may be stated much more simply and concisely by the formula $v = l \times w \times h$. This formula indicates that the volume v is equal to the product of the length l, the width w, and the height h of the prism. Where several letters are multiplied together in a formula, it is customary to omit the multiplication signs, the multiplication then being taken for granted. The foregoing formula, therefore, would ordinarily be written $v = lwh$. The multiplication sign must *not* be omitted between *numbers* that are to be multiplied together.

As may be seen from the example just given, a formula consists of two parts separated by the sign of equality. The letters or symbols denoting the quantities that are known are usually placed at the right of the equality sign, and the letter or symbol designating the value to be found is placed at the left of the equality sign. To apply a formula to the solution of an example a numerical value is substituted for each letter that denotes a known quantity, and the indicated mathematical operations are then performed. Care must be observed, in using a formula, to have all weights, dimensions, or other values expressed in the units required by the formula.

Letters with additional marks, such as C', a'', d_1, T_a, etc., are often found in formulas when similar quantities are to be represented by the same letter and yet to be distinguished from one another. The marks $'$ $''$ are termed *prime* and *second*, respectively, and the marks $_1$ and $_a$ are termed *subscripts* or *subs*. The four examples just given are read *large C prime*, *a second*, *d sub one*, and *large T sub a*. Parentheses

4

and brackets are used in formulas to indicate that the quantities enclosed by them are to be subjected to the same operation. The sign − before an expression in parentheses or brackets affects the entire expression, and if the parentheses or brackets are removed, the signs + and − within them must be interchanged; but if the sign + precedes the brackets, they may be removed without changing any signs. For example, the expression $212-(36+75-49)$ becomes, when the parentheses are removed, $212-36-75+49$; but the removal of the brackets from the expression $65+[20+9-14]$ gives $65+20+9-14$. The multiplication sign is ordinarily omitted before and after parentheses or brackets, and before the radical sign; thus, the expressions $36\times(18+22),[760-315]\times1.07$, and $.21\times\sqrt{140-27}$ would ordinarily be written $36(18+22)$, $[760-315]1.07$, and $.21\sqrt{140-27}$. The following examples will serve to illustrate the use of formulas.

EXAMPLE.—What is the volume of a block of cast iron 28 in. long, 15 in. wide, and 12 in. high?

SOLUTION.—Applying the formula previously given, and substituting 28 for l, 15 for w, and 12 for h, the volume is $v=28\times15\times12=5,040$ cu. in.

One of the most familiar formulas, to the operating engineer, is that used for finding the indicated horsepower of an engine. This formula, as usually stated, is

$$H=\frac{PLAN}{33,000}$$

in which $H=$ indicated horsepower;

   $P=$ mean effective pressure on piston, in pounds per square inch;
   $L=$ length of stroke, in feet;
   $A=$ area of piston, in square inches;
   $N=$ number of working strokes per minute.

The formula as stated may be used to find the horsepower of any engine, provided the values of the quantities denoted by P, L, A, and N are known. But sometimes it is desired to find some other quantity, as for example, the diameter of cylinder required to produce a certain horsepower, or the mean effective pressure necessary to produce the desired power.

To find these quantities, the order of the terms in the foregoing formula must be altered, bringing the values to be found to the left of the equality sign and the remainder to the right. This operation is called a *transformation* of a formula. The horsepower formula, transformed so as to give the values of the piston area and the mean effective pressure necessary to produce a desired horsepower, are

$$A = \frac{33,000H}{PLN}$$

and

$$P = \frac{33,000H}{LAN}$$

EXAMPLE.—Find the diameter of the piston of a steam engine that is designed to produce 40 H. P., if the stroke is 30 in., the mean effective pressure is 28 lb. per sq. in., and the speed is 75 R. P. *M.*

SOLUTION.—The length of stroke in feet is $L = 30 \div 12 = 2.5$ ft.; the mean effective pressure is $P = 28$ lb. per sq. in.; as there are two working strokes to each revolution, $N = 2 \times 75 = 150$; and $H = 40$. Substituting these values in the formula for the area of the piston,

$$A = \frac{33,000 \times 40}{28 \times 2.5 \times 150} = 125.7 \text{ sq. in.}$$

The diameter of a circle having an area of 125.7 sq. in. is about 12⅝ in. The required diameter of the piston, or of the cylinder, therefore, is 12⅝ in.

If a segment of a circle is not greater than a semicircle, its area may be found by the formula

$$A = \frac{\pi r^2 E}{360} - \frac{c}{2}(r - h),$$

in which A = area of segment;
$\pi = 3.1416$;
r = radius of circle;
E = angle, in degrees, obtained by drawing lines from the center to the extremities of arc of segment;
c = chord of segment;
h = height of segment.

EXAMPLE.—A segment of a circle having a radius of 7.5 in. is 1.91 in. high and its chord is 10 in. long. If the angle subtended by the chord is 83.46°, what is the area of the segment?

SOLUTION.—Substituting the given values in the foregoing formula,

$$A = \frac{3.1416 \times 7.5^2 \times 83.46}{360} - \frac{10}{2}(7.5 - 1.91)$$
$$= 40.97 - 27.95 = 13.02 \text{ sq. in., nearly.}$$

If the lengths of the sides of a triangle are known, the area may be found by the formula

$$A = \frac{b}{2}\sqrt{a^2 - \left(\frac{a^2 + b^2 - c^2}{2b}\right)^2},$$

in which A denotes the area of the triangle and a, b, and c denote the lengths of the three sides.

EXAMPLE.—What is the area of a triangle whose sides are 21 ft., 46 ft., and 50 ft. long?

SOLUTION.—In order to apply the formula, let a represent the side that is 21 ft. long; b, the side that is 50 ft. long; and c, the side that is 46 ft. long. Then, substituting in the formula,

$$A = \frac{50}{2}\sqrt{21^2 - \left(\frac{21^2 + 50^2 - 46^2}{2 \times 50}\right)^2}$$

$$= \frac{50}{2}\sqrt{441 - \left(\frac{441 + 2,500 - 2,116}{100}\right)^2} = 25\sqrt{441 - \left(\frac{825}{100}\right)^2}$$

$$= 25\sqrt{441 - 8.25^2} = 25\sqrt{441 - 68.0625} = 25\sqrt{372.9375}$$

$$= 25 \times 19.312 = 482.8 \text{ sq. ft., nearly.}$$

EXAMPLE.—When $x = 8$ and $y = 6$, what is the value of m in the following:

$$m = (x - y)(\sqrt[3]{x} + y) + \sqrt{\frac{(x^2 - y^2)^2}{4xy}}$$

SOLUTION.—Substituting,

$$m = (8 - 6)(\sqrt[3]{8} + 6) + \sqrt{\frac{(64 - 36)^2}{4 \times 8 \times 6}} = 2(2 + 6) + \sqrt{\frac{784}{192}}$$

$$= 16 + 2.02 = 18.02$$

MENSURATION

MEANINGS OF SYMBOLS

In the following formulas, the letters have the meanings here given, unless otherwise stated:

D = larger diameter;

d = smaller diameter;

R = radius corresponding to D;

r = radius corresponding to d;

p = perimeter or circumference;

C = area of convex surface = area of flat surface that can be rolled into the shape shown;

S = area of entire surface = C+area of the end or ends;

A = area of plane figure;

π = 3.1416, nearly = ratio of circumference of any circle to its diameter;

V = volume of solid.

The other letters used will be found on the illustrations.

TRIANGLES

Using letters to denote the angles,

$$D = B+C \qquad E + B + C = 180°$$
$$B = D-C \qquad E' + B + C = 180°$$
$$E' = E \qquad B' = B$$

For a right triangle, c being the hypotenuse,

$$c = \sqrt{a^2+b^2}$$
$$a = \sqrt{c^2-b^2}$$
$$b = \sqrt{c^2-a^2}$$

If c = length of side opposite an acute angle of an oblique triangle, and the distance e is known,

$$c = \sqrt{a^2+b^2-2be}$$
$$h = \sqrt{a^2-e^2}$$

If c = length of side opposite an obtuse angle of an oblique triangle,

$$c = \sqrt{a^2+b^2+2be}$$
$$h = \sqrt{a^2-e^2}$$

Any triangle inscribed in a semicircle is a right triangle, and

$$c:b = a:h$$

$$h = \frac{ab}{c} = \frac{ce}{a}$$

$$c:b+e = e:a = h:c$$

For any triangle,

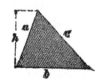

$$A = \frac{bh}{2} = \tfrac{1}{2}bh$$

$$A = \frac{b}{2}\sqrt{a^2 - \left(\frac{a^2+b^2-c^2}{2b}\right)^2}$$

Also, $\qquad A = \sqrt{s(s-a)(s-b)(s-c)}$

in which $s = \tfrac{1}{2}(a+b+c)$.

RECTANGLE AND PARALLELOGRAM

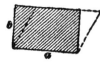

$$A = ab$$

$$A = b\sqrt{c^2-b^2}$$

TRAPEZOID

$$A = \tfrac{1}{2}h(a+b)$$

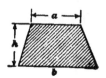

TRAPEZIUM

Divide the figure into two triangles and a trapezoid; then,

$$A = \tfrac{1}{2}bh' + \tfrac{1}{2}a(h'+h) + \tfrac{1}{2}ch$$

or, $A = \tfrac{1}{2}[bh' + ch + a(h'+h)]$

Or, divide into two triangles by drawing a diagonal. Considering the diagonal as the base of both triangles, call its length l, and call the altitudes of the triangles h_1 and h_2; then,

$$A = \tfrac{1}{2}l(h_1+h_2)$$

REGULAR POLYGONS

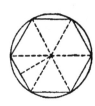

Divide the polygon into equal triangles and find the sum of the partial areas. Otherwise, square the length of one side and multiply by proper number from the following table:

Name	No. Sides	Multiplier	Name	No. Sides	Multiplier
Triangle......	3	.433	Heptagon....	7	3.634
Square.......	4	1.000	Octagon.....	8	4.828
Pentagon.....	5	1.720	Nonagon.....	9	6.182
Hexagon......	6	2.598	Decagon.....	10	7.694

IRREGULAR AREAS

Divide the area into trapezoids, triangles, parts of circles, etc., and find the sum of the partial areas.

If the figure is very irregular, the approximate area may be found as follows: Divide the figure into trapezoids by equidistant parallel lines b, c, d, etc., and measure the lengths of these lines. Then, calling a the first and n the last length, and y the width of strips,

$$A = y \left(\frac{a+n}{2} + b + c + \text{etc.} + m \right)$$

SECTOR

If l denotes the length of the arc, and $E*$ the angle in degrees and decimals of a degree,

$$l = \frac{Er}{57.296} = .0175 \ Er, \text{ nearly}$$

Then,

$$A = \tfrac{1}{2}lr$$

$$A = \frac{\pi r^2 E}{360} = .008727 r^2 E$$

CIRCLE

$$p = \pi d = 3.1416d$$

$$p = 2\pi r = 6.2832r$$

$$r = \frac{p}{2\pi} = \frac{p}{6.2832} = 1592p$$

---

*If the angle E is stated in degrees, minutes, and seconds, the minutes and seconds must be reduced to decimals of a degree. To do this, divide the number of minutes by 60 and the number of seconds by 3,600 and add the sum of the quotients to the number of degrees. Thus, $28° \ 42' \ 18'' = 28 + \frac{42}{60} + \frac{18}{3600} = 28 + .7 + .005 = 28.705°$.

$$p = 2\sqrt{\pi A} = 3.5449\sqrt{A}$$

$$p = \frac{2A}{r} = \frac{4A}{d}$$

$$d = \frac{p}{\pi} = \frac{p}{3.1416} = .3183p$$

$$d = 2\sqrt{\frac{A}{\pi}} = 1.1284\sqrt{A}$$

$$r = \sqrt{\frac{A}{\pi}} = .5642\sqrt{A}$$

$$A = \frac{\pi d^2}{4} = .7854d^2$$

$$A = \pi r^2 = 3.1416r^2$$

$$A = \frac{pr}{2} = \frac{pd}{4}$$

RING

$$A = \frac{\pi}{4}(D^2 - d^2)$$

SEGMENT

$$A = \tfrac{1}{2}[lr - c(r - h)]$$

$$A = \frac{\pi r^2 E}{360} - \frac{c}{2}(r - h)$$

$$l = \frac{\pi r E}{180} = .0175rE$$

$$E = \frac{180l}{\pi r} = 57.2956\frac{l}{r}$$

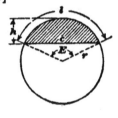

ELLIPSE

$$p^* = \pi \sqrt{\frac{D^2 + d^2}{2} - \frac{(D - d)^2}{8.8}}$$

$$A = \frac{\pi}{4}Dd = .7854Dd$$

CYLINDER

$$C = \pi dh$$

$$S = 2\pi rh + 2\pi r^2$$

$$= rdh + \frac{\pi}{2}d^2$$

$$V = \pi r^2 h = \frac{\pi}{4}d^2 h$$

$$V = \frac{p^2 h}{4\pi} = .0796p^2 h$$

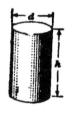

---

*The perimeter of an ellipse cannot be exactly determined without a very elaborate calculation, and this formula is merely an approximation giving close results.

FRUSTUM OF CYLINDER

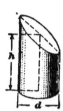

$h = \frac{1}{2}$ sum of greatest and least heights

$C = ph = \pi dh$

$S = \pi dh + \dfrac{\pi}{4}d^2 + \text{area of elliptic top}$

$V = Ah = \dfrac{\pi}{4}d^2 h$

PRISM OR PARALLELOPIPED

$C = Ph$
$S = Ph + 2A$
$V = Ah$

For prisms with regular polygons as bases, $P =$ length of one side $\times$ number of sides.

To obtain area of base, if it is a polygon, divide it into triangles and find sum of partial areas.

FRUSTUM OF PRISM

If a section perpendicular to the edges is a triangle, square, parallelogram, or *regular* polygon, $V = \dfrac{\text{sum of lengths of edges}}{\text{number of edges}} \times \text{area of right section.}$

SPHERE

$S = \pi d^2 = 4\pi r^2 = 12.5664 r^2$
$V = \pi d^3 = \frac{4}{3}\pi r^3 = .5236 d^3 = 4.1888 r^3$

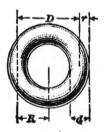

CIRCULAR RING

$D =$ mean diameter;
$R =$ mean radius.
$S = 4\pi^2 Rr = 9.8696 Dd$
$V = 2\pi^2 Rr^2 = 2.4674 Dd^2$

WEDGE

$V = \frac{1}{6}wh(a + b + c)$

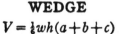

CIRCUMFERENCES AND AREAS OF CIRCLES FROM 1-64 TO 100

Diam.	Circum.	Area	Diam.	Circum.	Area
1/64	.0491	.0002	4	12.5664	12.5664
1/32	.0982	.0008	4 1/8	12.9591	13.3641
1/16	.1963	.0031	4 1/4	13.3518	14.1863
1/8	.3927	.0123	4 3/8	13.7445	15.0330
3/16	.5890	.0276	4 1/2	14.1372	15.9043
1/4	.7854	.0491	4 5/8	14.5299	16.8002
5/16	.9817	.0767	4 3/4	14.9226	17.7206
3/8	1.1781	.1104	4 7/8	15.3153	18.6555
5/16	1.3744	.1503	5	15.7080	19.6350
1/2	1.5708	.1963	5 1/8	16.1007	20.6290
9/16	1.7671	.2485	5 1/4	16.4934	21.6476
5/8	1.9635	.3068	5 3/8	16.8861	22.6907
11/16	2.1598	.3712	5 1/2	17.2788	23.7583
3/4	2.3562	.4418	5 5/8	17.6715	24.8505
13/16	2.5525	.5185	5 3/4	18.0642	25.9673
7/8	2.7489	.6013	5 7/8	18.4569	27.1086
15/16	2.9452	.6903	6	18.8496	28.2744
1	3.1416	.7854	6 1/8	19.2423	29.4648
1 1/8	3.5343	.9940	6 1/4	19.6350	30.6797
1 1/4	3.9270	1.2272	6 3/8	20.0277	31.9191
1 3/8	4.3197	1.4849	6 1/2	20.4204	33.1831
1 1/2	4.7124	1.7671	6 5/8	20.8131	34.4717
1 5/8	5.1051	2.0739	6 3/4	21.2058	35.7848
1 3/4	5.4978	2.4053	6 7/8	21.5985	37.1224
1 7/8	5.8905	2.7612	7	21.9912	38.4846
2	6.2832	3.1416	7 1/8	22.3839	39.8713
2 1/8	6.6759	3.5466	7 1/4	22.7766	41.2826
2 1/4	7.0686	3.9761	7 3/8	23.1693	42.7184
2 3/8	7.4613	4.4301	7 1/2	23.5620	44.1787
2 1/2	7.8540	4.9087	7 5/8	23.9547	45.6636
2 5/8	8.2467	5.4119	7 3/4	24.3474	47.1731
2 3/4	8.6394	5.9396	7 7/8	24.7401	48.7071
2 7/8	9.0321	6.4918	8	25.1328	50.2656
3	9.4248	7.0686	8 1/8	25.5255	51.8487
3 1/8	9.8175	7.6699	8 1/4	25.9182	53.4563
3 1/4	10.2102	8.2958	8 3/8	26.3109	55.0884
3 3/8	10.6029	8.9462	8 1/2	26.7036	56.7451
3 1/2	10.9956	9.6211	8 5/8	27.0963	58.4264
3 5/8	11.3883	10.3206	8 3/4	27.4890	60.1322
3 3/4	11.7810	11.0447	8 7/8	27.8817	61.8625
3 7/8	12.1737	11.7933	9	28.2744	63.6174

TABLE—(*Continued*)

Diam.	Circum.	Area	Diam.	Circum.	Area
9⅛	28.6671	65.3968	19½	61.2612	298.648
9¼	29.0598	67.2008	19¾	62.0466	306.355
9⅜	29.4525	69.0293	20	62.8320	314.160
9½	29.8452	70.8823	20¼	63.6174	322.063
9⅝	30.2379	72.7599	20½	64.4028	330.064
9¾	30.6306	74.6621	20¾	65.1882	338.164
9⅞	31.0233	76.589	21	65.9736	346.361
10	31.4160	78.540	21¼	66.7590	354.657
10¼	32.2014	82.516	21½	67.5444	363.051
10½	32.9868	86.590	21¾	68.3298	371.543
10¾	33.7722	90.763	22	69.1152	380.134
11	34.5576	95.033	22¼	69.9006	388.822
11¼	35.3430	99.402	22½	70.6860	397.609
11½	36.1284	103.869	22¾	71.4714	406.494
11¾	36.9138	108.434	23	72.2568	415.477
12	37.6992	113.098	23¼	73.0422	424.558
12¼	38.4846	117.859	23½	73.8276	433.737
12½	39.2700	122.719	23¾	74.6130	443.015
12¾	40.0554	127.677	24	75.3984	452.390
13	40.8408	132.733	24¼	76.1838	461.864
13¼	41.6262	137.887	24½	76.9692	471.436
13½	42.4116	143.139	24¾	77.7546	481.107
13¾	43.1970	148.490	25	78.5400	490.875
14	43.9824	153.938	25¼	79.3254	500.742
14¼	44.7678	159.485	25½	80.1108	510.706
14½	45.5532	165.130	25¾	80.8962	520.769
14¾	46.3386	170.874	26	81.6816	530.930
15	47.1240	176.715	26¼	82.4670	541.190
15¼	47.9094	182.655	26½	83.2524	551.547
15½	48.6948	188.692	26¾	84.0378	562.003
15¾	49.4802	194.828	27	84.8232	572.557
16	50.2656	201.062	27¼	85.6086	583.209
16¼	51.0510	207.395	27½	86.3940	593.959
16½	51.8364	213.825	27¾	87.1794	604.807
16¾	52.6218	220.354	28	87.9648	615.754
17	53.4072	226.981	28¼	88.7502	626.798
17¼	54.1926	233.706	28½	89.5356	637.941
17½	54.9780	240.529	28¾	90.3210	649.182
17¾	55.7634	247.450	29	91.1064	660.521
18	56.5488	254.470	29¼	91.8918	671.959
18¼	57.3342	261.587	29½	92.6772	683.494
18½	58.1196	268.803	29¾	93.4626	695.128
18¾	58.9050	276.117	30	94.2480	706.860
19	59.6904	283.529	30¼	95.0334	718.690
19¼	60.4758	291.040	30½	95.8188	730.618

MATHEMATICS
TABLE—(Continued)

Diam.	Circum.	Area	Diam.	Circum.	Area
30¾	96.6042	742.645	42	131.947	1,385.45
31	97.3896	754.769	42¼	132.733	1, 1.99
31¼	98.1750	766.992	42½	133.518	1, 8.63
31½	98.9604	779.313	42¾	134.303	1, 5.37
31¾	99.7458	791.732	43	135.089	1, 2.20
32	100.5312	804.250	43¼	135.874	1,409.14
32¼	101.3166	816.865	43½	136.660	1,486.17
32½	102.1020	829.579	43¾	137.445	1,503.30
32¾	102.8874	842.391	44	138.230	1,520.53
33	103.673	855.301	44¼	139.016	1,537.86
33¼	104.458	868.309	44½	139.801	1,555.29
33½	105.244	881.415	44¾	140.587	1,572.81
33¾	106.029	894.620	45	141.372	1,590.43
34	106.814	907.922	45¼	142.157	1,608.16
34¼	107.600	921.323	45½	142.943	1,625.97
34½	108.385	934.822	45¾	143.728	1,643.89
34¾	109.171	948.420	46	144.514	1,661.91
35	109.956	962.115	46¼	145.299	1,680.02
35¼	110.741	975.909	46½	146.084	1,698.23
35½	111.527	989.800	46¾	146.870	1,716.54
35¾	112.312	1,003.790	47	147.655	1,734.95
36	113.098	1,017.878	47¼	148.441	1,753.45
36¼	113.883	1,032.065	47½	149.226	1,772.06
36½	114.668	1,046.349	47¾	150.011	1,790.76
36¾	115.454	1,060.732	48	150.797	1,809.
37	116.239	1,075.213	48¼	151.582	1,828.
37¼	117.025	1,089.792	48½	152.368	1,847.
37½	117.810	1,104.469	48¾	153.153	1,866.
37¾	118.595	1,119.244	49	153.938	1,885.
38	119.381	1,134.118	49¼	154.724	1,905.
38¼	120.166	1,149.089	49½	155.509	1,924.
38½	120.952	1,164.159	49¾	156.295	1,943.
38¾	121.737	1,179.327	50	157.080	1,963.
39	122.522	1,194.593	50½	158.651	2,002.
39¼	123.308	1,209.958	51	160.222	2,042.
39½	124.093	1,225.420	51½	161.792	2,083.
39¾	124.879	1,240.981	52	163.363	2,123.
40	125.664	1,256.640	52½	164.934	2,164.
40¼	126.449	1,272.400	53	166.505	2,20 .56
40½	127.235	1,288.250	53½	168.076	2,246.
40¾	128.020	1,304.210	54	169.646	2,290.23
41	128.806	1,320.260	54½	171.217	2,332.83
41¼	129.591	1,336.410	55	172.788	2,375.83
41½	130.376	1,352.660	55½	174.359	2,419.23
41¾	131.162	1,369.000	56	175.930	2,463.01

TABLE—(*Continued*)

Diam.	Circum.	Area	Diam.	Circum.	Area
56½	177.500	2,507.19	78½	246.616	4,839.83
57	179.071	2,551.76	79	248.186	4,901.68
57½	180.642	2,596.73	79½	249.757	4,963.92
58	182.213	2,642.09	80	251.328	5,026.56
58½	183.784	2,687.84	80½	252.899	5,089.59
59	185.354	2,733.98	81	254.470	5,153.01
59½	186.925	2,780.51	81½	256.040	5,216.82
60	188.496	2,827.44	82	257.611	5,281.03
60½	190.067	2,874.76	82½	259.182	5,345.63
61	191.638	2,922.47	83	260.753	5,410.62
61½	193.208	2,9.0.58	83½	262.324	5,476.01
62	194.779	3,019.08	84	263.894	5,541.78
62½	196.350	3,067.97	84½	265.465	5,607.95
63	197.921	3,117.25	85	267.036	5,674.51
63½	199.492	3,166.93	85½	268.607	5,741.47
64	201.062	3,217.00	86	270.178	5,808.82
64½	202.633	3,267.46	86½	271.748	5,876.56
65	204.204	3,318.31	87	273.319	5,944.69
65½	205.775	3,369.56	87½	274.890	6,013.22
66	207.346	3,421.20	88	276.461	6,082.14
66½	208.916	3,473 24	88½	278.032	6,151.45
67	210.487	3,525.66	89	279.602	6,221.15
67½	212.058	3,578.48	89½	281.173	6,291.25
68	213.629	3,631.69	90	282.744	6,361.74
68½	215.200	3,685.29	90½	284.315	6,432.62
69	216.770	3,739.29	91	285.886	6,503.90
69½	218.341	3,793.68	91½	287.456	6,575.56
70	219.912	3,848.46	92	289.027	6 47.63
70½	221.483	3,903.63	92½	290.598	6, 20.08
71	223.054	3,959.20	93	292.169	6 92.92
71½	224.624	4,015.16	93½	293.740	6, 66.16
72	226.195	4,071.51	94	295.310	6,639.79
72½	227.766	4,128.26	94½	296.881	7,013.82
73	229.337	4,185.40	95	298.452	7,088.24
73½	230.908	4,242.93	95½	300.023	7,163.04
74	232.478	4,300.85	96	301.594	7,238.25
74½	234.049	4,359.17	96½	303.164	7,313.84
75	235.620	4,417.87	97	304.735	7,389.83
75½	237.191	4,476.98	97½	306.306	7,466.21
76	238.762	4,536.47	98	307.877	7,542.98
76½	240.332	4,596.36	98½	309.448	7,620.15
77	241.903	4,656.64	99	311.018	7,697.71
77½	243.474	4,717.31	99½	312.589	7,775.66
78	245.045	4,778.37	100	314.160	7,854.00

USEFUL TABLES

UNITS OF MEASUREMENT

LINEAR MEASURE

12 inches (in.)	= 1 foot	ft.
3 feet	= 1 yard	yd.
5½ yards	= 1 rod	rd.
40 rods	= 1 furlong	fur.
8 furlongs	= 1 mile	mi.

$$\text{mi.}\quad\text{fur.}\quad\text{rd.}\quad\text{yd.}\quad\text{ft.}\quad\text{in.}$$
$$1 = 8 = 320 = 1{,}760 = 5{,}280 = 63{,}360$$

There are various other units of length, such as the *league* = 3 mi.; the *nautical mile* = 6,080 ft.; the *fathom* = 6 ft.; the *hand* = 4 in.; the *span* = 9 in.; the *cubit* = 18 in.

SQUARE MEASURE

144 square inches (sq. in.)	= 1 square foot	sq. ft.
9 square feet	= 1 square yard	sq. yd.
30¼ square yards	= 1 square rod	sq. rd.
160 square rods	= 1 acre	A.
640 acres	= 1 square mile	sq. mi.

$$\text{sq. mi.}\quad\text{A.}\quad\text{sq. rd.}\quad\text{sq. yd.}\quad\text{sq. ft.}\quad\text{sq. in.}$$
$$1 = 640 = 102{,}400 = 3{,}097{,}600 = 27{,}878{,}400 = 4{,}014{,}489{,}600$$

CUBIC MEASURE

1,728 cubic inches (cu. in.)	= 1 cubic foot	cu. ft.
27 cubic feet	= 1 cubic yard	cu. yd.
128 cubic feet	= 1 cord	cd.
24¾ cubic feet	= 1 perch	P.

$$1 \text{ cu. yd.} = 27 \text{ cu. ft.} = 46{,}656 \text{ cu. in.}$$

MEASURES OF ANGLES OR ARCS

60 seconds (″)	= 1 minute	′
60 minutes	= 1 degree	°
90 degrees	= 1 rt. angle or quadrant	□
360 degrees	= 1 circle	

$$1 \text{ cir.} = 360° = 21{,}600' = 1{,}296{,}000''$$

A *quadrant* is one-fourth the circumference of a circle, or 90°; a *sextant* is one-sixth of a circle, or 60°. A *right angle* contains 90°. The unit of measurement is the degree, or $\frac{1}{360}$ of the circumference of a circle.

AVOIRDUPOIS WEIGHT

437½ grains (gr.)............... = 1 ounce.............. oz.
16   ounces................... = 1 pound.............. lb.
100   pounds................... = 1 hundredweight......cwt.
20   cwt., or 2,000 lb.......... = 1 ton................ T.
　　　　1 T. = 20 cwt. = 2,000 lb. = 32,000 oz. × 14,000,000 gr.
The avoirdupois pound contains 7,000 gr.

LONG-TON TABLE

16 ounces (oz.)............... = 1 pound.............. lb.
112 pounds................... = 1 hundredweight......cwt.
20 cwt., or 2,240 lb........... = 1 ton................ T.

TROY WEIGHT

24 grains (gr.)................ = 1 pennyweight........pwt.
20 pennyweights.............. = 1 ounce.............. oz.
12 ounces................... = 1 pound............. lb.
　　　　1 lb. = 12 oz. = 240 pwt. = 5,760 gr.

DRY MEASURE

2 pints (pt.).................. = 1 quart.............. qt.
8 quarts..................... = 1 peck................pk.
4 pecks..................... = 1 bushel.............bu.
　　　　1 bu. = 4 pk. = 32 qt. = 64 pt.
The U. S. struck bushel contains 2,150.42 cu. in. = 1.2444 cu. ft. By law, its dimensions are those of a cylinder 18½ in. in diameter and 8 in. deep. The heaped bushel is equal to 1¼ struck bushels, the cone being 6 in. high. The dry gallon contains 268.8 cu. in., being ⅛ struck bushel.

For approximations, the bushel may be taken as 1¼ cu. ft.; or 1 cu. ft. may be considered ⅘ bu.

The British bushel contains 2,218.19 cu. in. = 1.2837 cu. ft. = 1.032 U. S. bushels.

LIQUID MEASURE

4 gills (gi.)...............	=1 pint..............	pt.
2 pints..................	=1 quart.............	qt.
4 quarts................	=1 gallon............	gal.
31½ gallons.............	=1 barrel............	bbl.
2 barrels, or 63 gallons.......,	=1 hogshead..........	hhd.

1 hhd. = 2 bbl. = 63 gal. = 252 qt. = 504 pt. = 2,016 gi.

The U. S. gallon contains 231 cu. in. = .134 cu. ft., nearly, or 1 cu. ft. contains 7.481 gal.

When water is at its maximum density, 1 cu. ft. weighs 62.425 lb. and 1 gal. weighs 8.345 lb.

For approximations, 1 cu. ft. of water is considered equal to 7½ gal., and 1 gal. as weighing 8⅓ lb.

The British imperial gallon, both liquid and dry, contains 277.463 cu. in. = .16057 cu. ft., and is equivalent to the volume of 10 lb. of pure water at 62° F. To reduce British to U. S. liquid gallons, multiply by 1.2. Conversely, to convert U. S. into British liquid gallons, divide by 1.2; or, increase the number of gallons ⅙.

MEASURES OF UNITED STATES MONEY

10 mills (m.)...............	=1 cent...............	c.
10 cents.................	=1 dime...............	d.
10 dimes................	=1 dollar.............	$.
10 dollars...............	=1 eagle..............	E.

m.	c.	d.	$	E.
10 =	1			
100 =	10 =	1		
1,000 =	100 =	10 =	1	
10,000 =	1,000 =	100 =	10 =	1

The term *legal tender* is applied to money that may be legally offered in payment of debts. All gold coins are legal tender for their face value to any amount, provided their weight has not diminished more than $\frac{1}{100}$. Silver dollars also are legal tender to any amount, but silver coins of a lower denomination than $1 are legal tender only for sums not exceeding $10. Nickel and copper coins are legal tender for sums not exceeding 25c.

MEASURES OF TIME

60 seconds (sec.).............	= 1 minute.............	...min.
60 minutes.................	= 1 hour.............	hr.
24 hours..................	= 1 day..............	da.
7 days..................	= 1 week.............	wk.
4 weeks..................	= 1 month.............	mo.
12 months.................	= 1 year.............	yr.
100 years.................	= 1 century............	C.

$$yr. \quad wk. \quad da. \quad hr. \quad min. \quad sec.$$
$$1 = 52 = 365 = 8,765 = 525,948 = 31,556,936$$

DECIMAL EQUIVALENTS OF PARTS OF 1 IN.

1–64	.015625	17–64	.265625	33–64	.515625	49–64	.765625
1–32	.031250	9–32	.281250	17–32	.531250	25–32	.781250
3–64	.046875	19–64	.296875	35–64	.546875	51–64	.796875
1–16	.062500	5–16	.312500	9–16	.562500	13–16	.812500
5–64	.078125	21–64	.328125	37–64	.578125	53–64	.828125
3–32	.093750	11–32	.343750	19–32	.593750	27–32	.843750
7–64	.109375	23–64	.359375	39–64	.609375	55–64	.859375
1–8	.125000	3–8	.375000	5–8	.625000	7–8	.875000
9–64	.140625	25–64	.390625	41–64	.640625	57–64	.890625
5–32	.156250	13–32	.406250	21–32	.656250	29–32	.906250
11–64	.171875	27–64	.421875	43–64	.671875	59–64	.921875
3–16	.187500	7–16	.437500	11–16	.687500	15–16	.937500
13–64	.203125	29–64	.453125	45–64	.703125	61–64	.953125
7–32	.218750	15–32	.468750	23–32	.718750	31–32	.968750
15–64	.234375	31–64	.484375	47–64	.734375	63–64	.984375
1–4	.250000	1–2	.500000	3–4	.750000	1	1

WEIGHTS AND SIZES OF MATERIALS

SPECIFIC GRAVITY

The *specific gravity* of a substance is the ratio of the weight of any volume of the substance to the weight of an equal volume of water with solids and liquids, and air with gases. The weight of 1 cu. ft. of any solid or liquid, in pounds avoirdupois, is found by multiplying its specific gravity by 62.425. The weight, in pounds avoirdupois, of 1 cu. ft. of any gas at atmospheric pressure and at 32° F. is found by multiplying its specific gravity by .08073.

5

WEIGHTS OF VARIOUS SUBSTANCES

Metals	Weight p Cu. In.er Pound	Specific Gravity
Aluminum....................	.096	2.660
Antimony....................	.242	6.712
Bismuth.....................	.352	9.746
Brass, common..............	307	8.500
Copper, cast................	.314	8.700
Copper, rolled..............	321	8.878
Gold, pure cast.............	696	19.258
Iron, cast..................	260	7.207
Iron, wrought...............	.281	7.780
Lead, pure..................	.409	11.330
Mercury, at 60° F...........	491	13.580
Silver, pure................	.378	10.474
Steel, hard.................	.286	7.919
Steel, soft.................	.283	7.833
Tin.........................	.256	7.351
Zinc........................	.260	7.101

Stones and Earth	Weight per Cu. In. Pound	Specific Gravity
Asbestos....................	.1110	3 to 3.2
Brick.......................	.0723	2.000
Chalk.......................	.1006	2.784
Clay........................	.0686	1.900
Coal, anthracite............	.0592 / .0519	1.640 / 1.436
Coal, bituminous............	.0488	1.350
Earth, loose................	.0491	1.360
Emery.......................	.1450	4.000
Glass, flint................	.1260	3.500
Granite, Quincy.............	.0958	2.652
Gypsum, opaque..............	.0783	2.168
Limestone...................	.0980	2.700
Marble, common..............	.0970	2.686
Mica........................	.1012	2.800
Quartz......................	.0961	2.660
Salt, common................	.0769	2.130
Sand........................	.0957	2.650
Slate.......................	.1012	2.800
Soil, common................	.0717	1.984
Stone, common...............	.0910	2.520
Sulphur, native.............	.0734	2.033

TABLE—*(Continued)*

Dry Woods	Weight per Cu. In. Pound	Specific Gravity
Ash	.0305	.845
Beech	.0308	.852
Cedar, American	.0203	.561
Cork	.0090	.250
Ebony, American	.0441	1.220
Elm	.0202	.560
Lignum vitæ	.0481	1.330
Mahogany, Honduras	.0202	560
Maple	.0285	.790
Oak	.0343	.950
Pine, Southern	.0260	.720
Pine, White	.0144	.400
Poplar	.0138	.383
Spruce	.0181	.500

Liquids	Weight per Cu. In. Pound	Specific Gravity
Acid, nitric	.0440	1.217
Acid, sulphuric	.0665	1.841
Acid, muriatic, or hydrochloric	.0434	1.200
Alcohol, commercial	.0301	.833
Alcohol, pure	.0286	.792
Oil, linseed	.0340	.940
Oil, turpentine	.0314	.870
Water, distilled (62.425 lb. per cu. ft.)	.0361	1.000

Gases and Vapors At 32° and a Tension of 1 Atmosphere	Weight per Cu. Ft. Grains	Specific Gravity
Atmospheric air	565.11	1.0000
Ammonia gas	333.10	.5894
Carbonic acid	859.00	1.5201
Carbonic oxide	546.60	.9673
Hydrogen	39.10	.0692
Oxygen	624.80	1.1056
Sulphureted hydrogen	663.80	1.1747
Nitrogen	548.90	.9713
Steam at 212° F	275.80	.4880

STANDARD PIPE FOR STEAM, GAS, AND WATER

Nominal Inside Diameter Inches	Actual Inside Diameter Inches	Thickness Inch	Internal Area Square Inches	Threads Per Inch	Weight Per Foot Pounds
1/8	.27	.068	.06	27	.24
1/4	.36	.088	.10	18	.42
3/8	.49	.091	.19	18	.56
1/2	.62	.109	.30	14	.84
3/4	.82	.113	.53	14	1.12
1	1.05	.134	.86	11½	1.67
1¼	1.38	.140	1.50	11½	2.24
1½	1.61	.145	2.04	11½	2.68
2	2.07	.154	3.36	11½	3.61
2½	2.47	.204	4.78	8	5.74
3	3.07	.217	7.39	8	7.54
3½	3.55	.226	9.89	8	9.00
4	4.03	.237	12.73	8	10.66
4½	4.51	.246	15.96	8	12.49
5	5.05	.259	19.99	8	14.50
6	6.07	.280	28.89	8	18.76
7	7.02	.301	38.74	8	23.27
8	7.98	.322	50.02	8	28.18
9	8.94	.344	62.72	8	33.70
10	10.02	.366	78.82	8	40.00
11	11.00	.375	95.03	8	45.00
12	12.00	.375	113.10	8	49.00

WEIGHT OF SHEET IRON PER SQUARE FOOT

Number of Gauge	Thickness Inch	Black Iron Pounds	Number of Gauge	Thickness Inch	Black Iron Pounds	Galvanized Iron Pounds
1	.300	12.0	16	.065	2.6	3.0
2	.284	11.4	17	.058		
3	.259	10.4	18	.049	.	.
4	.238	9.5	19	.042	.	.
5	.220	8.8	20	.035	.	.
6	.203	8.1	21	.032	.	.
7	.180	7.2	22	.028	.	.
8	.165	6.6	23	.025	.	.
9	.148	5.9	24	.022	.	.
10	.134	5.4	25	.020	.	.
11	.120	4.8	26	.018	.	.
12	.109	4.4	27	.016	.	.
13	.095	3.8	28	.014	.	.
14	.083	3.3	29	.013	.	.
15	.072	2.9	30	.012	.	.

EXTRA-STRONG WROUGHT-IRON PIPE

Size of Pipe Inches	Nominal Inside Diameter Inches	Actual Outside Diameter Inches	Thickness Inch	Weight per Foot Pounds	Internal Area Square Inches
1/8	.205	.405	.100	.29	.03
1/4	.294	.540	.123	.54	07
3/8	.421	.675	.127	.74	14
1/2	.542	.840	.149	1.09	23
3/4	.736	1.050	.157	1.39	43
1	.951	1.315	.182	2.17	.71
1 1/4	1.272	1.660	.194	3.00	1.27
1 1/2	1.494	1.900	.203	3.63	1.75
2	1.933	2.375	.221	5.02	2.94
2 1/2	2.315	2.875	.280	7.67	4.21
3	2.892	3.500	.304	10.25	6.57
3 1/2	3.358	4.000	.321	12.47	8.86
4	3.818	4.500	.341	14.97	11.45
4 1/2	4.280	5.000	.360	18.22	14.39
5	4.813	5.563	.375	20.54	18.19
6	5.750	6.625	.437	28.58	25.98
7	6.625	7.625	.500	37.67	34.47
8	7.625	8.625	.500	43.00	45.66
9	8.625	9.625	.500	48.73	58.43
10	9.750	10.750	.500	54.74	74.66
11	10.750	11.750	.500	60.08	90.76
12	11.750	12.750	.500	65.42	108.43

WEIGHT OF IRON-PIPE SIZES OF SEAMLESS-DRAWN BRASS AND COPPER TUBES

Nominal Size Inches	Outside Diameter Inches	Inside Diameter Inches	Weight, in Pounds per Linear Foot	
			Brass	Copper
1/8	1 3/32	.	.30	.31
1/4			.43	.45
3/8			.58	.61
1/2		.	.80	.84
3/4	9/16	.	1.17	1.23
1	1 1/16	1.	1.67	1.75
1 1/4	1 5/8	1.	2.42	2.54
1 1/2	1 7/8	1.	2.92	3.07
2	2	2.	4.17	4.38
2 1/2	2 1/2	2.27	5.00	5.25
3	3	3.■■	8.00	8.40
4	4	4.02	12.00	12.00

WEIGHT OF ROUND AND SQUARE ROLLED IRON PER LINEAR FOOT

Side or Diameter Inches	Weight Pounds per Foot Round	Weight Pounds per Foot Square	Side or Diameter Inches	Weight Pounds per Foot Round	Weight Pounds per Foot Square
$\frac{1}{16}$	.010	.013	$3\frac{7}{8}$	39.864	50.756
$\frac{1}{8}$	.041	.053	4	42.464	54.084
$\frac{3}{16}$	.093	.118	$4\frac{1}{8}$	45.174	57.517
$\frac{1}{4}$	.165	.211	$4\frac{1}{4}$	47.952	61.055
$\frac{3}{8}$	.373	.475	$4\frac{3}{8}$	50.815	64.700
$\frac{1}{2}$	.663	.845	$4\frac{1}{2}$	53.760	68.448
$\frac{5}{8}$	1.043	1.320	$4\frac{5}{8}$	56.788	72.305
$\frac{3}{4}$	1.493	1.901	$4\frac{3}{4}$	59.900	76.264
$\frac{7}{8}$	2.032	2.588	$4\frac{7}{8}$	63.094	80.333
1	2.654	3.380	5	66.350	84.480
$1\frac{1}{8}$	3.359	4.278	$5\frac{1}{8}$	69.731	88.784
$1\frac{1}{4}$	4.147	5.280	$5\frac{1}{4}$	73.172	93.168
$1\frac{3}{8}$	5.019	6.390	$5\frac{3}{8}$	76.700	97.657
$1\frac{1}{2}$	5.972	7.604	$5\frac{1}{2}$	80.304	102.240
$1\frac{5}{8}$	7.010	8.926	$5\frac{5}{8}$	84.001	106.953
$1\frac{3}{4}$	8.128	10.352	$5\frac{3}{4}$	87.776	111.756
$1\frac{7}{8}$	9.333	11.883	$5\frac{7}{8}$	91.634	116.671
2	10.616	13.520	6	95.552	121.664
$2\frac{1}{8}$	11.988	15.263	$6\frac{1}{4}$	103.704	132.040
$2\frac{1}{4}$	13.440	17.112	$6\frac{1}{2}$	112.160	142.816
$2\frac{3}{8}$	14.975	19.066	$6\frac{3}{4}$	120.960	154.012
$2\frac{1}{2}$	16.588	21.120	7	130.048	165.632
$2\frac{5}{8}$	18.293	23.292	$7\frac{1}{4}$	139.544	177.672
$2\frac{3}{4}$	20.076	25.560	$7\frac{1}{2}$	149.328	190.136
$2\frac{7}{8}$	21.944	27.939	$7\frac{3}{4}$	159.456	203.024
3	23.888	30.416	8	169.856	216.336
$3\frac{1}{8}$	25.926	33.010	$8\frac{1}{4}$	180.696	230.068
$3\frac{1}{4}$	28.040	35.704	$8\frac{1}{2}$	191.808	244.220
$3\frac{3}{8}$	30.240	38.503	$8\frac{3}{4}$	203.260	258.800
$3\frac{1}{2}$	32.512	41.408	9	215.040	273.792
$3\frac{5}{8}$	34.886	44.418	$9\frac{1}{4}$	227.152	289.220
$3\frac{3}{4}$	37.332	47.534	$9\frac{1}{2}$	239.600	305.056

WEIGHT OF SHEET LEAD PER SQUARE FOOT

Thickness Inch	Weight Pounds	Thickness Inch	Weight Pounds	Thickness Inch	Weight Pounds
.017	1	.085	5	.152	9
.034	2	.101	6	.169	10
.051	3	.118	7	.186	11
.068	4	.135	8	.203	12

DIMENSIONS OF STANDARD AND EXTRA-HEAVY FLANGES

Size of Pipe Inches	Diameter of Flange Inches		No. of Holes		Diameter of Holes Inches		Diameter of Bolt Circle Inches		Thickness of Flange Inches	
	125 Lb.	250 Lb.	125 Lb.	250 Lb.	125 Lb.	250 Lb.	125 Lb.	250 Lb.	125 Lb.	250 Lb.
1	4	4½	4	4	9/16	5/8	3	3¼	7/16	11/16
1¼	4½	5	4	4	9/16	5/8	3⅜	3⅝	1/2	3/4
1½	5	6	4	4	5/8	3/4	3⅞	4½	9/16	13/16
2	6	6½	4	4	3/4	3/4	4¾	5	5/8	7/8
2½	7	7	4	4	3/4	3/4	5½	5⅞	11/16	15/16
3	7½	8½	4	8	3/4	3/4	6	6⅝	3/4	1
3½	8½	9	4	8	3/4	7/8	7	7¼	13/16	1⅛
4	9	10	8	8	3/4	7/8	7½	7⅞	15/16	1 3/16
4½	9¼	10½	8	8	3/4	7/8	7⅞	8⅝	15/16	1¼
5	10	11	8	8	7/8	7/8	8½	9¼	15/16	1 5/16
6	11	12½	8	12	7/8	7/8	9½	10⅝	1	1⅜
7	12½	14	8	12	7/8	7/8	10⅝	11¾	1 1/16	1 7/16
8	13½	15	12	12	7/8	7/8	11¾	13	1⅛	1½
9	15	16½	12	12	7/8	1	13¼	14	1 3/16	1 9/16
10	16	18½	12	16	1	1	14¼	15¼	1¼	1 11/16
12	19	20½	12	16	1	1⅛	17	17	1⅜	1¾
14	21	23	14	20	1⅛	1¼	18¾	20	1⅜	2
15	22¼	25	16	20	1⅛	1¼	20	21¼	1 7/16	2⅛
16	23½	26	16	20	1⅛	1⅜	21¼	22½	1 9/16	2¼
18	25	28½	16	24	1¼	1⅜	22¾	24¼	1 11/16	2⅜
20	27½	31	20	24	1¼	1½	25	27	1¾	2½
22	29½	33	20	28	1¼	1⅝	27¼	29¼	1 13/16	2¾
24	32	36	20	28	1¼	1¾	29½	32	1⅞	3

DIMENSIONS OF PIPE FLANGES

The table on page 51 gives the dimensions of standard and extra-heavy pipe flanges, the former being the pressure up to 125 lb. per sq. in., and the latter for pressures above 125 and

STANDARD AND EXTRA-GAUGE STEEL BOILER TUBES

Out-side Diameter Inches	Standard Thickness		Nominal Weight per Foot Pounds				
	Nearest Birm. Wire Gauge	Inch	Standard Thickness	One Extra Wire Gauge	Two Extra Wire Gauges	Three Extra Wire Gauges	Four Extra Wire Gauges
1	13	.095	.90	1.04	1.13	1.24	1.35
1¼	13	.095	1.15	1.33	1.45	1.60	1.74
1½	13	.095	1.40	1.62	1.77	1.96	2.14
1¾	13	.095	1.66	1.91	2.09	2.31	2.53
2	13	.095	1.91	2.20	2.41	2.67	2.93
2¼	13	.095	2.16	2.49	2.73	3.03	3.32
2½	12	.109	2.75	3.05	3.39	3.72	4.12
2¾	12	.109	3.04	3.37	3.74	4.11	4.56
3	12	.109	3.3	3.69	4.10	4.51	5.00
3¼	11	.120	3.9	4.46	4.90·	5.44	5.90
3½	11	.120	4.2	4.82	5.30·	5.88	6.38
3¾	11	.120	4.6	5.18	5.69	6.32	6.86
4	10	.134	5.4	6.09	6.76	7.34	8.23
4½	10	.134	6.1	6.8	7.64	8.31	9.32
5	9	.148	7.58	8.58	9.27	10.40	11.23
6	8	.165	10.16	11.19	12.57	13.58	14.65
7	8	.165	11.90	13.11	14.74	15.93	17.19
8	8	.165	13.65	15.04	16.91	18.28	19.73
9	7	.180	16.76	19.07	20.63	22.27	24.18
10	6	.203	21.00	22.98	24.82	26.95	29.47
11	5	.220	25.00	27.36	29.71	32.51	34.29
12	4½	.229	28.50	31.19	34.01	36.52	39.92
13	4	.238	32.06	35.25	38.57	40.70	45.98

not over 250 lb. per sq. in. The dimensions of flanges and fittings were arranged by joint committees from the American Society of Mechanical Engineers and the National Association of Master Steam and Hot-Water Fitters, and after revision and amplification were adopted by the latter body and officially

STANDARD LAP-WELDED CHARCOAL-IRON BOILER TUBES

Diameter Inches		Thickness inch	Circumference Inches		Area Square Inches		Length of Tube in Feet per Square Foot of Surface		Weight per Foot Pounds
Outside	Inside		Outside	Inside	Outside	Inside	Outside	Inside	
1	.810	.095	3.142	2.545	.785	.515	3.820	4.479	.90
1¼	1.060	.035	3.927	3.330	1.227	.882	3.056	3.604	1.15
1½	1.310	.095	4.712	4.115	1.767	1.348	2.547	2.916	1.40
1¾	1.560	.095	5.498	4.901	2.405	1.911	2.183	2.448	1.65
2	1.810	.095	6.283	5.686	3.142	2.573	1.910	2.110	1.91
2¼	2.060	.095	7.069	6.472	3.976	3.333	1.698	1.854	2.16
2½	2.282	.09	7.854	7.169	4.909	4.090	1.528	1.674	2.75
2¾	2.532	.109	8.639	7.955	5.940	5.035	1.389	1.508	3.04
3	2.782	.109	9.425	8.740	7.069	6.079	1.273	1.373	3.33
3¼	3.010	.120	10.210	9.456	8.296	7.116	1.175	1.269	3.96
3½	3.260	.120	10.996	10.242	9.621	8.347	1.091	1.172	4.28
3¾	3.510	.120	11.781	11.027	11.045	9.676	1.019	1.088	4.60
4	3.732	.134	12.566	11.724	12.566	10.939	.955	1.024	5.47
4½	4.232	.134	14.137	13.295	15.904	14.066	.849	.903	6.17
5	4.704	.148	15.708	14.778	19.635	17.379	.764	.812	7.58
6	5.670	.165	18.850	17.813	28.274	25.250	.637	.674	10.16
7	6.670	.165	21.991	20.954	38.485	34.942	.546	.573	11.90
8	7.670	.165	25.133	24.096	50.266	46.204	.477	.498	13.65
9	8.640	.180	28.274	27.143	63.617	58.630	.424	.442	16.76
10	9.594	.203	31.416	30.141	78.540	72.292	.382	.398	21.00

recommended by the former. Also, the American Society of
Heating and Ventilating Engineers recommended the values
for the use of its members. The table gives the dimensions
of flanges, but not of fittings. Another series of values is the
Manufacturers' 1912 Schedule, which was adopted July 10,
1912, to take effect Oct. 1, 1912. The latter schedule agrees
with the table on page 51, so far as standard flanges are concerned, but has slightly different values for extra-heavy flanges
for pipes over 8 in. in diameter.

CHEMISTRY AND HEAT

CHEMISTRY

Divisions of **Matter.**—*Matter* is anything that occupies
space, and it exists in three states, namely, solid, liquid, and
gaseous. It is made up of molecules and atoms. A *molecule*
is the smallest portion of matter into which a body can be
divided and exist without changing its nature. An *atom* is
an indivisible portion of matter, or the smaller particle produced by dividing a molecule. A molecule is simply a group
of two or more atoms that are held together by their natural
attraction for one another.

Elements **and** Compounds.—Every body, or every portion
of matter, is either an element, a compound, or a mixture.
Iron, silver, sulphur, and oxygen are elements; wood, coal,
salt, and water are compounds; and atmospheric air is a mixture. An *element* is a substance that cannot be divided or
broken up into other substances; thus, if a piece of silver is
divided and subdivided, each particle will still be silver. A
compound is a substance that can be divided into other substances; thus, if an electric current is passed through water,
the water is decomposed into two gases, hydrogen and oxygen.
A *mixture* is simply a combination of elements or compounds
in which each preserves its own nature; thus, air is a mixture of oxygen and nitrogen and minute proportions of
other gases.

Symbols and Formulas.—Each of the known elements is designated by a *symbol*, which is a letter or a pair of letters, oftentimes the initial letter of the name of the element. Thus, O is the symbol for oxygen, H for hydrogen, C for carbon, and so on. When two or more elements unite to form a compound, the symbols representing the elements in the compound may be connected in a *formula*, which will show how the atoms of the elements combined to form the compound. Thus, H and O are the symbols of the elements hydrogen and oxygen. When these two gases unite in certain proportions, they form water, a compound whose formula is H_2O. This formula indicates, first, that water is composed of hydrogen, H, and oxygen, O; and second, it indicates, by the subscript 2, that 2 atoms of hydrogen unite with 1 atom of oxygen to form 1

ATOMIC WEIGHTS OF ELEMENTS

Name of Element	Symbol	Atomic Weight	Name of Element	Symbol	Atomic Weight
Calcium.....	Ca	40.10	Nitrogen....	N	14.04
Carbon......	C	12.00	Oxygen.....	O	16.00
Chlorine.....	Cl	35.45	Potassium ..	K	39.15
Hydrogen ...	H	1.00	Sodium.....	Na	23.05
Magnesium..	Mg	24.36	Sulphur....	S	32.06

molecule of water. The formula for carbon dioxide is CO_2, which indicates that a molecule of carbon dioxide consists of 1 atom of carbon, C, and 2 atoms of oxygen, O.

Atomic Weight **and Molecular Weight.**—The ratio between the weight of an atom of any element and the weight of an atom of hydrogen is termed the *atomic weight* of that element. The symbols and the atomic weights of a number of the elements most commonly met with in steam engineering are given in the accompanying table. By the aid of the atomic weights, the composition of any substance, by weight, may be determined. For example, water contains 2 atoms of hydrogen and 1 atom of oxygen. Multiplying the number of atoms of each by the atomic weight, it is seen that there are $2 \times 1 = 2$ parts of hydrogen, by weight, and, $1 \times 16 = 16$ parts of oxygen,

COMMON NAMES OF CHEMICAL COMPOUNDS

Chemical Compound	Common Name	Formula
Ammonium chloride..	Sal ammoniac.	NH_4Cl
Ammonium hydrate..	Liquor ammonia.....	NH_4OH
Calcium hydroxide....	Slaked lime... ...	$Ca(OH)_2$
Calcium chloro-hypo-chlorite............	Bleaching powder, or chloride of lime....	$Ca(ClO)Cl$
Calcium oxide........	Quicklime..........	CaO
Calcium sulphate.....	Plaster of Paris......	$2CaSO_4 \cdot H_2O$
Hydrochloric acid	Muriatic acid........	HCl
Magnesium sulphate..	Epsom salts........	$MgSO_4$
Nitric acid..........	Aquafortis..........	HNO_3
Potassium hydrate ...	Caustic potash.......	KOH
Potassium nitrate.....	Niter, or saltpeter....	KNO_3
Sodium carbonate....	Washing soda........	$Na_2CO_3 \cdot 10H_2O$
Sodium carbonate....	Soda ash............	Na_2CO_3
Sodium chloride ...	Common salt........	$NaCl$
Sodium hydrate ...	Caustic soda.........	$NaOH$
Sulphuric acid........	Oil of vitriol.........	H_2SO_4

by weight, and the molecule contains $2+16=18$ parts. The weight of the molecule of a substance is termed the *molecular weight*.

HEAT

TEMPERATURE

Thermometric **Scales.**—Three different scales are used for designating temperatures, namely, the Fahrenheit scale, the centigrade scale, and the Réaumur scale. The last-named is little used, however. Temperatures on these scales are usually indicated by the abbreviations F. or Fahr., C. or Cent., and R. or Réau., respectively. There are three standard points on each scale, namely, the boiling point of water, at sea level, the melting point of ice, and the absolute zero of temperature. The last indicates the point of complete absence of heat. On the Fahrenheit scale, the melting point of ice is marked 32 and the boiling point of water is marked 212, and the intervening space is divided into 180 equal parts, called *degrees*.

On the centigrade scale, the melting point of ice is zero, and the boiling point of water is 100, and there are 100 divisions, or degrees, between these points. On the Réaumur scale, zero marks the melting point of ice, and 80 the boiling point of water.

Absolute Zero.—It has been found by experiment that all perfect gases will expand $\frac{1}{460}$ of their volume when heated from zero to 1° above it. It is inferred, therefore, that the

CENTIGRADE AND FAHRENHEIT DEGREES

Deg. C.	Deg. F.	Deg. C.	Deg. F.	Deg. C.	Deg. F.	Deg. C.	Deg. F.
0	32.0	26	78.8	51	123.8	76	168.8
1	33.8	27	80.6	52	125.6	77	170.6
2	35.6	28	82.4	53	127.4	78	172.4
3	37.4	29	84.2	54	129.2	79	174.2
4	39.2	30	86.0	55	131.0	80	176.0
5	41.0	31	87.8	56	132.8	81	177.8
6	42.8	32	89.6	57	134.6	82	179.6
7	44.6	33	91.4	58	136.4	83	181.4
8	46.4	34	93.2	59	138.2	84	183.2
9	48.2	35	95.0	60	140.0	85	185.0
10	50.0	36	96.8	61	141.8	86	186.8
11	51.8	37	98.6	62	143.6	87	188.6
12	53.6	38	100.4	63	145.4	88	190.4
13	55.4	39	102.2	64	147.2	89	192.2
14	57.2	40	104.0	65	149.0	90	194.0
15	59.0	41	105.8	66	150.8	91	195.8
16	60.8	42	107.6	67	152.6	92	197.6
17	62.6	43	109.4	68	154.4	93	199.4
18	64.4	44	111.2	69	156.2	94	201.2
19	66.2	45	113.0	70	158.0	95	203.0
20	68.0	46	114.8	71	159.8	96	204.8
21	69.8	47	116.6	72	161.6	97	206.6
22	71.6	48	118.4	73	163.4	98	208.4
23	73.4	49	120.2	74	165.2	99	210.2
24	75.2	50	122.0	75	167.0	100	212.0
25	77.0						

ultimate limit of contraction will be found at 460° below zero on the Fahrenheit scale, and that at this point all motion of the molecules ceases. This point is called the *absolute zero*, and temperatures measured therefrom are called *absolute temperatures.*

The temperature that is indicated by the Fahrenheit thermometer may be converted into absolute temperature by adding it to 460°. Thus, a temperature of 85° by the Fahrenheit thermometer corresponds to the absolute temperature of $85+460=545°$. On the centigrade scale the absolute zero is $273\frac{1}{3}°$ below the zero point. On the Réaumur scale it is $218\frac{2}{3}°$ below zero. When the thermometer indicates temperatures below the zero point of its graduation, the indicated temperature must be subtracted from 460, $273\frac{1}{3}$, or $218\frac{2}{3}$, respectively, to find the absolute temperature; that is, absolute zero is $-460°$ F., $-273\frac{1}{3}°$ C., and $-218\frac{2}{3}°$ R.

Conversion of Temperatures.—A degree on the Fahrenheit scale is equal to $\frac{100}{180}=\frac{5}{9}$ of a degree centigrade and to $\frac{80}{180}=\frac{4}{9}$ of a degree Réaumur. Temperatures according to any one of these scales, therefore, may be converted into the corresponding temperatures on the other scales by using the following simple formulas:

Temp. F.$=\frac{9}{5}$ Temp. C.$+32°=\frac{9}{4}$ Temp. R.$+32°$.

Temp. C.$=\frac{5}{9}$ (Temp. F.$-32°$)$=\frac{5}{4}$ Temp. R.

Temp. R.$=\frac{4}{9}$ (Temp. F.$-32°$)$=\frac{4}{5}$ Temp. C.

The table on page 57 shows the equivalents of centigrade temperatures on the Fahrenheit scale.

COEFFICIENTS OF LINEAR EXPANSION

The *coefficient of expansion* of a body is its expansion per degree rise of temperature. The coefficient of surface expansion is double, and that of cubic expansion three times the coefficient of linear expansion. The table on page 59 shows the coefficient of linear expansion for various substances.

For example, a 30-ft. steel rail in warming from 20° F. below zero to 100° F. will expand $(20+100)\times.00000599\times30\times12$ $=.2588$ in.

MEASUREMENT OF HEAT

British Thermal Unit.—The unit most commonly used for the measurement of heat is the British Thermal Unit, abbreviated B. T. U. This is the amount of heat required to raise the temperature of 1 lb. of pure water 1° F. at or near the temperature 39.1° F., which is the point of maximum density of water. Heat is a form of energy, and it may be transformed into other forms of energy. The equivalent of 1 B. T. U. in foot-pounds is 778 ft.-lb., and this value, 778 ft.-lb., is termed the *mechanical equivalent of heat*. It is the number of foot-pounds of mechanical energy that would be produced by transforming 1 B. T. U. without any losses.

COEFFICIENTS OF EXPANSION FOR VARIOUS SUBSTANCES

Substance		Coefficient of Linear Expansion in Inches per Degree F.
Aluminum		.00001140
Brass		00001040
Brick		00000306
Cement and Concrete	{ from	.00000550
	to	.00000780
Copper		00000961
Glass	{ from	.00000399
	to	.00000521
Gold		00000841
Granite		00000460
Iron, cast		00000587
Iron, wrought		00000677
Lead		00001580
Marble		00000400
Masonry	{ from	.00000206
	to	.00000490
Mercury		00003334
Platinum		00000494
Porcelain		00000200
Sandstone	{ from	.00000400
	to	.00000670
Steel, untempered		00000599
Steel, tempered		00000702
Tin		00001160
Wood, pine		00000276
Zinc		.00001634

Latent Heat.—The heat expended in changing a body from the solid to the liquid state, or from the liquid to the gaseous state, without change of temperature, is called its *latent heat*.

The temperature at which a body changes from a solid to a liquid state is called its *temperature of fusion*, or its *fusing point*; and the number of B. T. U. required to effect this change in a body weighing 1 lb. is called its *latent heat of fusion*. The temperature at which a body changes from a liquid state to a vapor or a gas is called its *temperature of vaporization;* and the heat required to effect this change in 1 lb. of the liquid is called its *latent heat of vaporization.*

When a vapor changes back to a liquid, it is said to condense, and when a liquid changes back to a solid, it is said to freeze; in either case, an amount of heat, equal to the latent heat of vaporization or of fusion, as the case may be, must be abstracted from, or given up, by the body.

TEMPERATURES AND LATENT HEATS OF FUSION AND OF VAPORIZATION

Substance	Temperature of Fusion Deg. F.	Latent Heat of Fusion B. T. U.	Temperature of Vaporization Deg. F.	Latent Heat of Vaporization B. T. U.
Aluminum........	1,160.0	51.4		
Carbon..........	Infusible			
Copper..........	1.930.0			
Ice..............	32.0	144.0	212	
Iron, cast........	2,192.0	233.0		
Iron, wrought.....	2,912.0			
Lead............	626.0	9.7		
Mercury.........	−37.8	6.0	662	157
Platinum........	3,227.0	4		
Steel............	2,520.0			
Sulphur.........	239.0	16.9	824	
Tin.............	446.0	25.7		
Zinc............	680.0	50.6	1,900	493

The accompanying table shows the latent heats of fusion and of vaporization for 1 lb. of various substances, they having first been raised to the temperature at which the change takes place, and the pressure being one atmosphere, or 14.7 lb. per sq. in. The temperature of vaporization in the table is the boiling point of the liquid under the ordinary atmospheric pressure of 14.7 lb. per sq. in.

Specific Heat.—The *specific heat* of a body is the ratio between the quantity of heat required to warm that body 1° and the quantity of heat required to warm an equal weight of water 1°. The number of B. T. U. required in order to raise or required to be abstracted in order to lower, the temperature of a body a certain number of degrees may be found by the formula $Q = Ws\ (t_1 - t_2)$,

in which
Q = number of B. T. U. required;
W = weight of body, in pounds;
s = specific heat of body;
t_1 = higher temperature, in degrees F.;
t_2 = lower temperature, in degrees F.

The specific heats of various solids, liquids, and gases are given in the accompanying tables.

SPECIFIC HEATS OF SOLIDS AND LIQUIDS

Substance	Specific Heat	Substance	Specific Heat
Aluminum	.2143	Lead, melted	.0402
Ashes, average	.2100	Mercury	.0333
Brass	.0939	Platinum	.0324
Charcoal	.2410	Steel	.1170
Copper	.0951	Sulphur	.2026
Glass	.1937	Sulphur, melted	.2340
Ice	.5040	Tin	.0562
Iron, cast	.1298	Tin, melted.	.0637
Iron, wrought	.1138	Water	1.0000
Lead	.0314	Zinc	.0956

SPECIFIC HEATS OF GASES

Name of Gas	Specific Heat	
	Constant Pressure	Constant Volume
Air	.23751	.16902
Carbon dioxide	.21700	.15350
Carbon monoxide	.24500	.17580
Hydrogen	3.40900	2.41226
Nitrogen	.24380	.17273
Oxygen	.21751	.15507

MECHANICS

WORK AND POWER

Work is the overcoming of resistance through a distance. The unit of work is the *foot-pound;* that is, it equals 1 lb. raised vertically 1 ft. The amount of work done is equal to the resistance in pounds multiplied by the distance in feet through which it is overcome. If a body is lifted, the resistance is the weight or the overcoming of the attraction of gravity, the work done being the weight in pounds multiplied by the height of the lift in feet. If a body moves in a horizontal direction, the work done is the friction overcome, or the force needed to move a resistant body or combination of bodies, multiplied by the distance moved through.

It must always be kept in mind that motion in itself is not work and that the mere application of a force also is not work; a force must act through a distance overcoming resistance in order that work be done.

Power is the rate of doing work, or the quantity of work done in unit time. The ordinary unit of mechanical power is the *horsepower*, which is equivalent to 33,000 ft.-lb. per min., or 550 ft.-lb. per sec. The term horsepower is commonly abbreviated H. P.

The work necessary to be done in raising a body weighing W lb. through a height of h ft. equals $W h$ ft.-lb. The total work that any moving body is capable of doing in being brought to rest equals its *kinetic energy*, or $\dfrac{Wv^2}{2g}$, in which v is the velocity of the body, in feet per second, and $g = 32.16$.

The kinetic energy of a 200,000-lb. train running at 40 mi. per hr., or 58.7 ft. per sec., is $200,000 \times 58.7^2 \div (2 \times 32.16)$ $= 10,714,220$ ft.-lb.; the retarding force necessary to stop the train within 2,000 ft. is $10,714,220 \div 2,000 = 5,357.1$ lb., and the average power required to stop the train in $\frac{1}{2}$ min. is $10,714,220 \div \frac{1}{2} = 21,428,440$ ft.-lb. per min., or $21,428,440 \div 33,000 = 649.3$ H. P.

CENTRIFUGAL FORCE

If a ball is fastened to a string and is whirled so as to give it a circular motion, there will be a pull of greater or less amount on the string, according as the speed of the ball increases or decreases. If the string is cut while the ball is in motion, the ball will fly off, away from the center of the circle in which it is whirling. The force that acts to draw a body away from the center around which it revolves is termed the *centrifugal force* of the body. It may be found by the formula

$$F = .00034 \ WRN^2,$$

in which F = centrifugal force, in pounds;

$\quad\quad W$ = weight of revolving body, in pounds;

$\quad\quad R$ = radius, in feet, of circle in which the center of gravity of the revolving body moves;

$\quad\quad N$ = revolutions per minute of revolving body.

MACHINE ELEMENTS

LEVERS

A *lever* is a bar that may be turned about a pivot, or point, as shown in Figs. 1, 2, and 3. In each case, the object W to

FIG. 1　　　　FIG. 2　　　　FIG. 3

be lifted is called the *load*, or *weight;* the *force* that accomplishes the lifting is represented by F; and the *fulcrum*, or pivot, is indicated by C. The distance l from the weight to the fulcrum is termed the *weight arm* and the distance L from the force to the fulcrum is termed the *force arm*. The distance between the force and the weight is denoted by a. Whenever the force F is just great enough to balance the load lifted, it will be found that the force times the length of the force arm is equal to the weight times the length of the weight arm. That is, in each of the forms of levers shown in Figs. 1 to 3, $F:W=l:L$, or $FL=Wl$. From this it follows that $F=\dfrac{Wl}{L}$ and $W=\dfrac{FL}{l}$ If the force and the weight are known, and it is desired to calculate the lengths of the weight and force arms so that the lever may balance, the following formulas may be used:

For the style of lever shown in Fig. 1,
$$l=\frac{Fa}{W+F} \text{ and } L=\frac{Wa}{W+F}$$
For the style of lever shown in Fig. 2,
$$l=\frac{Fa}{W-F} \text{ and } L=\frac{Wa}{W-F}$$
For the style of lever shown in Fig. 3,
$$l=\frac{Fa}{F-W} \text{ and } L=\frac{Wa}{F-W}$$

FIXED AND MOVABLE PULLEYS

A *pulley* consists of a grooved wheel on an axle held in a frame, or *block*, and is very useful in moving or hoisting loads. A *fixed pulley* is one whose block is not movable. Such a pulley is shown in Fig. 1. A rope passes over the pulley and carries the load W at one end, the hoisting force F being applied

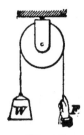 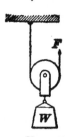 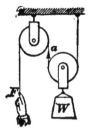 

FIG. 1   FIG. 2   FIG. 3   FIG. 4

at the other end of the rope. Neglecting the friction of the pulley in its block, the force required to lift a load is equal to the load; that is, $F=W$.

A *movable pulley* is one whose block is movable, such as that shown in Fig. 2. In this case, one end of the rope is fastened to an overhead support and the hoisting force is applied at the other end in an upward direction. With this arrangement, a pull of 1 lb. at F will lift a weight of 2 lb. at W. If the free end of the rope is carried up over a fixed pulley, as shown in Fig. 3, the effect will still be the same; that is, a weight of 2 lb. can be raised by a force F of 1 lb.; that is, in Figs. 2 and 3, $F = \frac{1}{2}W$.

A double movable pulley is shown in Fig. 4; that is, the movable block attached to the load W carries two pulleys. A similar pulley is fixed to an overhead support, and one end of the rope is attached to it, the other end being carried around the several pulleys as shown. With this arrangement, a force F of 1 lb. is sufficient to raise a weight W of 4 lb.; that is, $F = \frac{1}{4}W$.

Combination of Pulleys.—By increasing the number of fixed and movable pulleys over which the hoisting rope or chain passes, the force required to lift a given load may be lessened. In Fig. 5 is shown a quadruple movable pulley, or one in which the movable block has four pulleys, and a force of 1 lb. at F will lift a weight of 8 lb. at W. The following general rule applies to any combination of pulleys:

Rule.—*When a single rope is used with a combination of pulleys, a pull on the free end will balance a load on the movable block as many times as great as the pull as there are parts of the rope supporting the movable block*

For instance, in Figs. 2 and 3, there are two parts of the rope supporting the movable block, and the load balanced by the pull F is twice as great as the pull F. In Fig. 4 there are four parts of the rope supporting the movable block, so that $W = 4F$, or $F = \frac{1}{4}W$. In Fig. 5, $W = 8F$, as there are eight parts of the rope supporting the movable block.

Differential Pulley.—The arrangement shown in Fig. 6 is known as a differential pulley. The two fixed pulleys have different radii r and R. The hoisting chain passes around the larger of the fixed pulleys, then around the movable pulley carrying the weight W, and then around the smaller fixed pulley. The chain is joined at its ends, thus forming a continuous piece. The two fixed pulleys are fastened together so that they must turn as one piece on their axle. When a pull P is exerted on a chain, the fixed pulleys are rotated. The chain winds up on one side of the large pulley and unwinds, at a slower rate, from the small pulley. As a result,

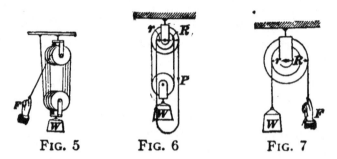

FIG. 5   FIG. 6   FIG. 7

the load W is lifted slightly at each turn of the fixed pulleys. The weight W that can be lifted by a force P is found by the formula

$$W = \frac{2PR}{R-r},$$

the letters R and r designating the radii of the large and small fixed pulleys, respectively.

Wheel and Axle.—The device known as the wheel and axle consists of two cylinders of different sizes connected rigidly, so that they turn together. The rope carrying the load to be raised is fastened to the smaller cylinder and the rope on which the pulling force is exerted is wound on the larger cylinder, as shown in Fig. 7. The pull F causes the larger cylinder to turn, rotating the smaller cylinder at the same rate. The rope attached to the weight W is thus caused to wind on the smaller cylinder, and the load is raised. If R and r represent, respectively, the radii of the large and small cylinders, the force F required to lift a weight W, and the weight W that may be lifted by a force F, are given by the formulas

$$F = \frac{Wr}{R} \quad \text{and} \quad W = \frac{FR}{r}$$

BELT PULLEYS

Solid and Split Pulleys.—Besides being used with ropes or chains for the hoisting of loads, pulleys are extensively employed with belts for transmitting power. Belt pulleys may be divided into two classes, namely, solid pulleys and split pulleys. A *solid pulley* is one in which the arms, hub, and rim are cast in one solid piece, as shown in Fig. 1. A *split pulley* is one that is cast in halves that are afterwards bolted together, as shown in Fig. 2. The latter style of pulley is more readily placed on or removed from a shaft than is the solid pulley. Pulleys are generally cast in halves or parts when they are more than 6 ft. in diameter. This is done on account of the shrinkage strains in large pulley castings, which renders the pulleys liable to crack as a result of unequal cooling of the metal.

 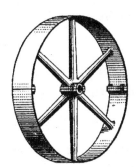

FIG. 1   FIG. 2

Wooden Pulleys.—Although most belt pulleys are made of cast iron, wrought iron, and steel, wooden pulleys have come into extensive use. These are built of segments of wood securely glued together, maple being the wood ordinarily used. It is possible to procure wooden split pulleys that are fitted with removable bushings, thus allowing the same pulley to be adapted readily to shafting of different diameters. Wooden pulleys are somewhat lighter than cast-iron pulleys for the same service.

Driving and Driven Pulleys.—The pulley that imparts motion to the belt is called the *driving pulley*, or the *driver*, and the one that receives motion from the belt is called the *driven pulley*, or simply the driven. When two pulleys are connected by a belt, the speeds at which the pulleys run are inversely proportional to their diameters. Thus, if two pulleys have diameters of 12 in. and 24 in., the speed of the smaller is to the speed of the larger as 24 to 12, or as 2 to 1.

The speed of a pulley or of a shaft is usually stated in revolutions per minute, abbreviated R. P. M.

Diameter and Speed of Driver.—It often becomes necessary to calculate the size or the speed of a pulley that drives or is being driven by a machine.

Let D = diameter of driving pulley, in inches;

d = diameter of driven pulley, in inches;

N = number of R. P. M. of driving pulley;

n = number of R. P. M. of driven pulley.

Then, if the diameter of the driven and the required speeds of both pulleys are given, the diameter of the driver may be found by the formula

$$D = \frac{dn}{N} \qquad (1)$$

If the speed of the driver is to be found, it is necessary to use the formula

$$N = \frac{dn}{D} \qquad (2)$$

EXAMPLE.—A 12-in. pulley on a certain machine is to run at 160 R. P. M. and is to be driven by belt from a pulley on the shaft of an engine that makes 96 R. P. *M.* What must be the diameter of the pulley on the engine shaft?

SOLUTION.—Substituting in formula 1,

$$D = \frac{12 \times 160}{96} = 20 \text{ in.}$$

Diameter and Speed of Driven.—If the diameter of the driving pulley and the desired speeds of both pulleys are known, the required diameter of the driven pulley may be found by the formula

$$d = \frac{DN}{n} \qquad (1)$$

If the speed of the driven pulley is to be found, it is necessary to use the formula

$$n = \frac{DN}{d} \qquad (2)$$

EXAMPLE 1.—A 30-in. pulley on a line shaft running at 120 R. P. M. is to drive a pulley on a machine at 300 R. P. *M.* What must be the diameter of the pulley on the machine?

SOLUTION.—Substituting in formula 1,

$$d = \frac{30 \times 120}{300} = 12 \text{ in.}$$

EXAMPLE 2.—A driving pulley 48 in. in diameter makes 175 R. P. M. and is connected by belt to a driven pulley 14 in. in diameter. What is the speed of the driven pulley?

SOLUTION.—Substituting in formula 2,

$$n = \frac{48 \times 175}{14} = 600 \text{ R. P. M.}$$

BELTING

A *belt* is a flexible band by which motion is transmitted from one pulley to another. The materials most commonly used for belts are leather, cotton, and rubber, although thin, flat bands of steel are coming into use. Leather belts are usually made either single or double. A *single belt* is one composed of a single thickness of leather, and a *double belt* is one composed of two thicknesses of leather cemented and riveted together throughout the whole length of the belt. Still heavier belts, consisting of three or four thicknesses of leather, and known as *triple* or *quadruple belts*, are sometimes made for heavy drives. Cotton belts are made up of a number of layers, or plies, sewed together and treated with a water-proofing substance. They are termed *two-ply, three-ply*, etc., according to the number of plies they contain. Four-ply cotton belting is usually considered equal to single leather belting. Rubber belts are particularly adapted for use in damp or wet places. They withstand changes of temperature without injury, are durable, and are claimed to be less liable to slip than are leather belts.

Sag of Belts.—The distance between pulley centers depends on the size of the pulleys and of the belt; it should be great enough so that the belt will run with a slight sag and a gently undulating motion, but not great enough to cause excessive sag and an unsteady flapping motion of the belt. In general, the centers of small pulleys carrying light narrow belts should be about 15 ft. apart and the belt sag 1½ to 2 in.; for large pulleys and heavy belts the distance should be 20 to 30 ft. and the sag 2½ to 5 in. Loose-running belts will last much longer than tight ones, and will be less likely to cause heating and wear of bearings.

Speed of Belts.—The higher the speed of a belt, the less may be its width to transmit a given horsepower; consequently, it follows that a belt should be run at as high a speed as conditions will permit. The greatest allowable speed for a belt joined by lacing is about 3,500 ft. per min., for ordinary single and double leather belts. For belts joined by cementing, when the joint has about the same strength as the solid belt, the velocity may be as high as 5,000 ft. per min. Higher speeds than these have been used, but there is little to be gained by exceeding about 4,800 ft. per min. In choosing a proper belt speed, due regard must be paid to commercial conditions. Although a high speed of the belt means a narrow and cheaper belt, the increased cost of the larger pulleys that may be required may offset the gain due to the high speed of the belt, at least so far as the first cost is concerned. The speed of a belt, in feet per minute, may be found by multiplying the number of revolutions per minute of the pulley by 3.1416 times the diameter of the pulley, in inches, and dividing the product by 12.

Horsepower of Belts.—The pull on a belt is greatest on the tight, or driving, side, and least on the slack side. The difference between the tensions, or pulls, in these two sides is called the *effective pull*. The effective pull that may be allowed per inch of width for single leather belts with different arcs of contact is given in the accompanying table. The *arc of contact* is the portion of the circumference of the smaller pulley that is covered by the belt. The horsepower that can be transmitted by a single leather belt may be found by the formula

$$H = \frac{CWV}{33,000}, \qquad (1)$$

in which H = horsepower of belt;

　　　　C = effective pull, taken from table;

　　　　W = width of belt, in inches;

　　　　V = speed of belt, in feet per minute.

If it is desired to find the width of single belt required to transmit a given horsepower, the formula becomes

$$W = \frac{33,000\,H}{CV} \qquad (2)$$

EXAMPLE 1.—What horsepower can be transmitted by a single leather belt 4 in. wide running at a speed of 2,500 ft. per min., if the belt covers one-third of the circumference of the small pulley?

SOLUTION.—The fraction of the circumference covered by the belt is $\frac{1}{3} = .333$. From the table, the allowable effective pull corresponding to this value is 28.8. Substituting in formula 1,

$$H = \frac{28.8 \times 4 \times 2,500}{33,000} = 8.7 \text{ H. P.}$$

EXAMPLE 2.—A single leather belt is to run at a speed of 3,000 ft. per min. and is to transmit 18 H. P. Find the width of the belt, if the arc of contact is 150°

SOLUTION.—The effective pull corresponding to an arc of contact of 150°, from the table, is 33.8. Substituting in formula 2,

$$W = \frac{33,000 \times 18}{33.8 \times 3,000} = 5.9 \text{ in.}$$

A 6-in. belt would be used.

ALLOWABLE EFFECTIVE PULL

Arc of Contact		Allowable Effective Pull
Degrees	Fraction of Circumference	Pounds per Inch of Width
90	250	23.0
112½	.312	27.4
120	.333	28.8
135 .	.375	31.3
150	.417	33.8
157½	437	34.9
180 or over	.500 or over	38.1

The horsepower of a double leather belt may be taken as 1¾ times that of a single leather belt of the same width running under the same conditions.   Accordingly, the width of a double leather belt required for any service is only $\frac{7}{10}$ that of a single belt for the same service.

Lacing of Belts.—A very satisfactory way of lacing belts less than 3 in. wide is shown in Fig. 1, in which A is the outside of the belt and B is the side that runs against the face of the pulley.   The ends of the belt are first cut square, and then holes are punched in the ends, in corresponding positions opposite one another.   The number of holes in each row should always be odd, in the style of lacing shown, using 3 holes in belts up to 2 in. wide and 5 holes in belts between 2 and 3 in. wide.   The lacing is first drawn through one of the middle holes from the under side. or pulley side, as at 1. Then it is drawn across the upper side and is passed down through 2, across under the belt to 3, up through 3, across and down again through 2, back under the belt and up through 3 again, then across and down through 4 and finally up through 5, where a barb is cut in the edge of the lacing to prevent it from pulling out.   This completes the lacing of one half. The other end of the lacing is then carried through the holes in the other half, in the same order.

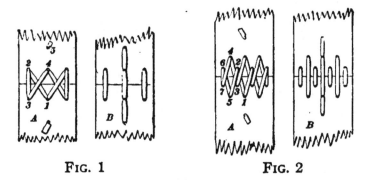

FIG. 1   FIG. 2

For belts wider than 3 in., the lacing shown in Fig. 2 may be used.   In this case, there are two rows of holes in each end of the belt to be joined.   The row nearer the end of the belt should have one more hole than the row farther away.   For belts up to $4\frac{1}{2}$ in. wide, use 3 holes in the first row and 2 holes in the second row.   For belts up to 6 in. wide, use 4 and 3 holes, respectively.   For wider belts, make the total number of holes in both rows either one or two more than the number of inches of width of the belt, the object being to get an odd total number of holes. For example, a 10-in. belt would have $10+1=11$ holes, and a 13-in. belt would have $13+2=15$ holes. The outside holes of the first row should not be nearer the side edges of the belt than ¾ in. and not nearer the joint edge than ⅛ in. The second row should be at least 1¾ in. from the joint edge. In Fig. 2, *A* is the outside face and *B* the face next the pulley. The lacing is first drawn up through 1 from the pulley side, and then is carried through 2, 3, 4, 5, 6, 7, 6, 7, 4, 5, 2, 3, 8, and out at 9 to be fastened. The other end of the lacing is used on the other half of the belt in the same way.

HYDRAULICS

Pressure Due to Head of Water.— *Hydraulics* treats of liquids in motion. When a tank is filled with water, its bottom and its sides are subjected to pressure due to the weight of the water. The distance from a point at any given level in the tank to the surface of the water, measured vertically, is termed the *head* of water for that level. The pressure exerted by the water is directly proportional to the head. A cubic foot of water weighs approximately 62.5 lb., so that, if a tank 1 sq. ft. in cross-section were filled with water to a depth of 1 ft., it would contain 1 cu. ft., or 62.5 lb., of water. The pressure on the bottom, having an area of 1 sq. ft., would be 62.5 lb., or the weight of 1 cu. ft. of water; consequently, the pressure per square inch would be $62.5 = 144 = .434$ lb. In other words, a column of water 1 ft. deep, or having a head of 1 ft., exerts a pressure of .434 lb. per sq. in. Knowing this fact, the pressure due to any given head of water may be found by the formula

$$p = .434\ h,$$

in which p = pressure in pounds per square inch;

h = head of water, in feet.

If it is desired to find the head of water necessary to produce a certain pressure, the formula becomes

$$h = \frac{p}{.434} = 2.304\ p$$

Velocity of Flow.—If a hole is made in the side of a tank filled with water, the water will issue therefrom with a velocity depending on the head of water above the opening.

When water flows in a pipe, a ditch, or a channel of any kind, the velocity is not the same at all points, because the cross-section of the channel is not the same at all points, and also because of friction. In such cases, the mean velocity is taken in all calculations. The *mean velocity* is that velocity which, being multiplied by the area of the cross-section of the stream, will equal the total quantity discharged.

Flow of Water in Pipes.—For straight cylindrical pipes of uniform diameter, the mean velocity of discharge may be calculated by the formula

$$V_m = 2.315 \sqrt{\frac{hd}{fl + 125d}}, \qquad (1)$$

in which V_m = mean velocity of discharge in feet per second;

h = total head in feet = vertical distance between the level of water in reservoir and the point of discharge;

l = length of pipe, in feet;

d = diameter of pipe, in inches;

f = coefficient of friction.

The head is always taken as the vertical distance between the point of discharge and the level of the water at the source, or point from which it is taken, and is always measured in feet. It matters not how long the pipe is, whether vertical or inclined, whether straight or curved, nor whether any part of the pipe goes below the level of the point of discharge or not; the head is always measured as stated above.

EXAMPLE.—What is the mean velocity of efflux from a 6-in. pipe, 5,780 ft. long, if the head is 170 ft.? Take f = .021.

SOLUTION.—Substituting in formula 1,

$$V_m = 2.315 \sqrt{\frac{170 \times 6}{.021 \times 5,780 + (.125 \times 6)}} = 6.69 \text{ ft. per sec.}$$

When the pipe is very long compared with the diameter, as in the foregoing example, use may be made of the formula

$$V_m = 2.315 \sqrt{\frac{hd}{fl}}, \qquad (2)$$

in which the letters have the same meaning as in the preceding formula. This formula may be used when the length of the pipe exceeds 10,000 times its diameter.

The actual head necessary to produce a certain velocity V_m may be calculated by the formula

$$h = \frac{fl V_m^2}{5.36\,d} + .0233\ V_m^2 \qquad (3)$$

If the head, the length of the pipe, and the diameter of the pipe are given, to find the discharge, use the formula

$$Q = .09445\ d^2 \sqrt{\frac{h\,d}{fl + .125\,d}}, \qquad (4)$$

in which Q = discharge in U. S. gallons per second.

To find the value of f, calculate V_m by formula 2, assuming that $f = .025$, and get the final value of f from the following table:

V_m	f	V_m	f	V_m	f
1	.0686	.7	.0349	2	.0265
.2	.0527	.8	.0336	3	.0243
.3	.0457	.9	.0325	4	.0230
.4	.0415	1	.0315	6	.0214
.5	.0387	1¼	.0297	8	.0205
.6	.0365	1½	.0284	12	.0193

EXAMPLE.—The length of a pipe is 6,270 ft., its diameter is 8 in. and the total head at the point of discharge is 215 ft. How many gallons are discharged per minute?

SOLUTION.— $V_m = 2.315 \sqrt{\dfrac{215 \times 8}{.025 \times 6,270}} = 7.67$ ft. per sec., nearly. Using the value of $f = .0205$ for $V_m = 8$ (see table),

$$Q = .09455 \times 8^2 \sqrt{\frac{215 \times 8}{.0205 \times 6,270 + (.125 \times 8)}} = 22.03 \text{ gal. per sec.}$$

$= 22.03 \times 60 = 1,321.8$ gal. per min.

If it is desired to find the head necessary to give a discharge of a certain number of gallons per second through a pipe whose length and diameter are known, calculate the mean velocity of efflux by using the formula

$$V_m = \frac{24.51Q}{d^2}; \quad (5)$$

find the value of f from the table, corresponding to this value of V_m, and substitute these values of f and V_m in the formula for the head.

EXAMPLE.—A 4-in. pipe, 2,000 ft. long, is to discharge 24,000 gal. of water per hr.; what head is necessary?

SOLUTION.— $Q = \dfrac{24,000}{60 \times 60} = 6\frac{2}{3}$ gal. per sec. $V_m = \dfrac{24.51 \times 6\frac{2}{3}}{4^2}$

$= 10.2$ ft. per sec. From the table, $f = .0205$ for $V_m = 8$, and 0193 for $V_m = 12$; assume that $f = .02$ for $V_m = 10.2$. Then

$$h = \frac{.02 \times 2,000 \times 10.2^2}{5.36 \times 4} + .0233 \times 10.2^2 = 196.53 \text{ ft.}$$

COMBUSTION AND FUELS

COMBUSTION

Nature of Combustion.—*Combustion* is the very rapid chemical combination of two or more elements, accompanied by the production of light and heat. The atoms of some of the elements have a very great affinity or attraction for those of other elements, and when they combine they rush together with such rapidity and force that heat and light are produced. Oxygen, for example, has a great attraction for nearly all the other elements. For carbon, oxygen has a particular liking, and whenever these two elements come into contact at a sufficiently high temperature, they combine with great rapidity. The combustion of coal in the furnace of a boiler is of this nature. The temperature of the furnace is raised by kindling the fire, and then the carbon of the coal begins to combine with oxygen taken from the air.

Products of Combustion.—When carbon and oxygen combine they form CO_2, or carbon dioxide; when hydrogen and oxygen combine they form water, H_2O. These are called the *products of combustion.* The oxygen required for combustion is usually obtained from the air, which is a mixture composed of approximately 23 parts of oxygen and 77 parts of nitrogen by weight. The nitrogen that enters the furnace with the oxygen takes no part in the combustion, but passes through the furnace and up the chimney without any change in its nature.

Air Required for Combustion.—When carbon is burned to carbon dioxide, CO_2, 1 atom of carbon unites with 2 atoms of oxygen. Carbon has an atomic weight of 12 and oxygen has an atomic weight of 16, so that the molecular weight of CO_2 is $(1 \times 12) + (2 \times 16) = 44$; hence CO_2 is composed of $12 \div 44 = 27.27$ per cent. of carbon and $32 \div 44 = 72.73\%$ of oxygen. To burn a pound of carbon to CO_2, therefore, requires $32 \div 12 = 2\frac{2}{3}$ lb. of oxygen. If the oxygen is taken from the air, it will take $2\frac{2}{3} \div .23 = 11.6$ lb. of air to supply the $2\frac{2}{3}$ lb. of oxygen. This is because only 23% of air is oxygen. The combustion of a pound of carbon to CO_2 may be represented as follows:

Mixture in Pounds		*Elements in Pounds*		*Products in Pounds*	
Carbon,	1.0	Carbon,	1.00 ⎫	= Carbon dioxide,	3.67
Air,	11.6 =	{ Oxygen,	2.67 ⎭		
		⎩ Nitrogen,	8.93	Nitrogen,	8.93
	12.6		12.60		12.60

That is, 1 lb. of carbon requires 11.6 lb. of air for complete combustion. Of this air, 2.67 lb. is oxygen which combines with the pound of carbon, forming 3.67 lb. of carbon dioxide. The 8.93 lb. of nitrogen contained in the air passes off with the CO_2 as a product of combustion.

Take, next, the complete combustion of 1 lb. of hydrogen. The product of the combustion is water, H_2O. It has been shown that H_2O is composed by weight of 2 parts hydrogen to 16 parts oxygen. Hence 1 lb. of H requires $16 \div 2 = 8$ lb. of O to unite with it. The air required to furnish 8 lb. of O is $8 \div .23 = 34.8$ lb. The process of combustion is, therefore, as follows:

Mixture in Pounds		*Elements in Pounds*		*Products in Pounds*	
Hydrogen,	1.0	Hydrogen,	1.0 $\Big\}$ = Water,		9.0
Air,	34.8 = $\Big\{$	Oxygen,	8.0		
		Nitrogen,	26.8	Nitrogen,	26.8
	·35.8		35.8		35.8

Incomplete Combustion.—There is one other case that may occur; the combustion of carbon may not be complete. If insufficient air or oxygen is supplied to the burning carbon, it is possible for the carbon and oxygen to form another gas, carbon monoxide, CO, instead of carbon dioxide, CO_2. The combustion of 1 lb. of carbon to form CO, of course, requires only one-half the oxygen that would be necessary to form CO_2. This is because in CO gas 1 atom of carbon seizes 1 atom of oxygen instead of 2. To burn 1 lb. of carbon to CO_2 requires 11.6 lb. of air. To burn it to CO would, therefore, require but 5.8 lb. of air.

Calorific Value of Fuels.—The amount of heat, in B. T. U., developed by the complete combustion of 1 lb. of a fuel is termed the *calorific value* of that fuel; it is also sometimes called the *heat value* or the *heat of combustion*. It may be determined most accurately by burning a known weight of the fuel with oxygen in an instrument known as a *calorimeter*. The gases resulting from the combustion are passed through a known weight of water and give up their heat to the water. By noting the rise of temperature of the water, it is possible to calculate the amount of heat absorbed, and thus to determine the heat that would be produced by the combustion of 1 lb. of the fuel. The calorific values of the elements most commonly found in fuels are as follows:

B. *T. U. per Lb.*

Hydrogen, burned to water, H_2O 62,000
Carbon, burned to CO_2 14,600
Carbon, burned to CO 4,400
Sulphur, burned to SO_2 4,000

If the various percentages, by weight of the elements, composing a fuel are known, the approximate calorific value of that fuel may easily be calculated by the formula

$$X = 14,600C + 62,000 \left(H - \frac{O}{8} \right) + 4,000S,$$

in which X = calorific value of fuel, in B. T. U. per pound;

C = percentage of carbon, expressed as a decimal;

H = percentage of hydrogen, expressed as a decimal;

O = percentage of oxygen, expressed as a decimal;

S = percentage of sulphur, expressed as a decimal.

EXAMPLE.—A coal contains 85% of carbon, 4% of oxygen, 6% of hydrogen, 1% of sulphur, and 4% of ash. What is the heat of combustion per pound?

SOLUTION.—Applying the formula, $X = 14,600 \times .85 + 62,000$

$$\left(.06 - \frac{.04}{8}\right) + 4,000 \times .01 = 15,860 \text{ B. T. U.}$$

FUELS

SOLID FUELS

Fuels for Steam Making.—The fuels used in the generation of steam are chiefly coal, coke, wood, petroleum, and natural gas. Other fuels, such as the waste gases from blast furnaces, straw, bagasse, dried tan bark, green slabs, sawdust, peat, etc., are also used. All these fuels are composed either of carbon alone or carbon in combination with hydrogen, oxygen, sulphur, and non-combustible substances.

Classes of Coal.—Coal is the fuel most extensively used in steam-plant work. Its different varieties may be classed in four main groups, namely, anthracite, semianthracite, semibituminous, and bituminous coal.

Anthracite Coal.—Anthracite coal contains from 92.31 to 100% of fixed carbon and from 0 to 7.69% of volatile hydrocarbons. It is rather hard to ignite and requires a strong draft to burn it. It is quite hard and shiny; in color it is a grayish black. It burns with almost no smoke, and this fact gives it a peculiar value in places where smoke is objectionable. Anthracite coal is known to the trade by different names, according to the size into which the lumps are broken. These names, with the generally accepted dimensions of the screens over and through which the lumps of coal will pass, are as follows: *Culm* passes through $\frac{3}{16}$-in. round mesh. *Rice* passes over $\frac{3}{16}$-in. mesh and through $\frac{1}{4}$-in. square mesh. *Buckwheat*

No. 2 passes over ¼-in. mesh and through ⁵⁄₁₆-in. mesh. *Buck-wheat No. 1* passes over ⁵⁄₁₆-in. mesh and through ½-in. square mesh. *Pea* passes over ½-in. mesh and through ¾-in. square mesh. *Chestnut* passes over ¾-in. mesh and through 1⅜-in. square mesh. *Stove* passes over 1⅜-in. mesh and through 2-in. square mesh. *Egg* passes over 2-in. mesh and through 2¼-in. square mesh. *Broken* passes over 2¼-in. mesh and through 3½-in. square mesh. *Steamboat* passes over 3½-in. mesh and out of screen. *Lump* passes over bars set from 3½ to 5 in. apart.

Semianthracite Coal. — Semianthracite coal contains from 87.5 to 92.31% of fixed carbon and from 7.69 to 12.5% of volatile hydrocarbons. It kindles easily and burns more freely than the true anthracite coal; hence, it is highly esteemed as a fuel. It crumbles readily and may be distinguished from anthracite coal by the fact that when just fractured it will soil the hand, while anthracite will not do so. It burns with very little smoke. Semianthracite coal is broken into different sizes for the market; these sizes are the same and are known by the same trade names as the corresponding sizes of anthracite coal.

Semibituminous Coal.—Semibituminous coal contains from 75 to 87.5% of fixed carbon and from 12.5 to 25% of volatile hydrocarbons. It differs from semianthracite coal only in having a smaller percentage of fixed carbon and more volatile hydrocarbons. Its physical properties are practically the same, and since it burns without the smoke and soot emitted by bituminous coal, it is a valuable steam fuel. Semibituminous and bituminous coals are known to the trade by the following names: *Lump coal* includes all coal passing over screen bars 1½ in. apart. *Nut coal* passes over bars ¾ in. apart and through bars 1½ in. apart. *Pea coal* passes over bars ⅜ in. apart and through bars ¾ in. apart. *Slack* includes all coal passing through bars ⅜ in. apart.

Bituminous Coal.—Bituminous coal contains from 0 to 75% of fixed carbon and from 25 to 100% of volatile hydrocarbons. It may be divided into three classes, whose names and characteristics are as follows: *Caking coal* is the name given to coals that, when burned in the furnace, swell and fuse together, forming a spongy mass that may cover the whole surface of the grate. These coals are difficult to burn, because the fusing prevents the air from passing freely through the bed of burning fuel. When caking coals are burned, the spongy mass must be frequently broken up with the slice bar, in order to admit the air needed for its combustion. *Free-burning coal* is a class of bituminous coal that is often called *non-caking coal* from the fact that it has no tendency to fuse together when burned in a furnace. *Cannel coal* is a grade of bituminous coal that is very rich in hydrocarbons. The large percentage of volatile matter makes it valuable for gas making, but it is little used for the generation of steam, except near the places where it is mined.

Lignite.—Lignite, or brown coal, contains from 30 to 60% of carbon, a small quantity of hydrocarbons, and a large amount of oxygen. It occupies a position between peat and bituminous coal, being probably of a later origin than the latter. It has an uneven fracture and a dull luster. Its value as a steam fuel is limited, since it will easily break in transportation. Exposure to the weather causes it to absorb moisture rapidly, and it will then crumble quite readily. It is non-caking and yields but a moderate heat, and is in this respect inferior to even the poorer grades of bituminous coal.

Miscellaneous Fuels.—*Coke* is made from bituminous coal by driving off the volatile matter. It consists of from 88 to 95% of carbon, $\frac{1}{4}$ to 2% of sulphur, and from 4 to 12% of ash. It is little used for steam-boiler fuel.

Wood is used for fuel in localities where it is plentiful. It contains from 20 to 50% of moisture when cut, and this percentage is not reduced much below 20% by drying. Wood has a calorific value of 6,000 to 7,000 B. T. U. per lb.

Peat consists of vegetable matter that is partly carbonized and is found at the surface of the earth. It contains from 75 to 80% water when cut, and must be dried before it can be used as fuel.

Bagasse is the refuse left after the juice has been extracted from the sugar cane by means of the rolls. It is used to some extent in tropical and semitropical countries. Naturally, its use is limited to the places where the sugar cane is grown.

Dried tan bark, straw, slabs, and sawdust being refuse, their use is local and usually confined to tanneries, planing and sawmills, and threshing outfits.

The Babcock & Wilcox Company state that on the average 1 lb. of good bituminous coal may be considered as the equivalent of 2 lb. of dry peat, $2\frac{1}{2}$ lb. of dry wood, $2\frac{1}{4}$ to 3 lb. of dry tan bark or sun-dried bagasse; 3 lb. of cotton stalks, $3\frac{3}{4}$ lb. of straw, 6 lb. of wet bagasse, and from 6 to 8 lb. of wet tan bark.

LIQUID FUEL

Nature of Petroleum.—The fuel most extensively employed in the generation of steam is coal, the most valuable of the solid fuels.   In some parts of the world, however, it has been found convenient and economical to use liquid fuel.   This is obtained chiefly from *petroleum*, which is a natural oil obtained from the earth.   In its original state it is usually of a dark-green color when viewed in the sunlight; but when held up to the light, so that the light passes through it, it has a reddish-brown color. The appearance of the oil will vary somewhat, depending on the locality from which it is derived.   In some cases it is almost as clear and colorless as water, and in other cases it is black; but American petroleum is commonly brown or reddish-brown with a green luster.

Composition of Crude Oil.—Petroleum in the form in which it issues from the earth is known as *crude oil*.   It usually contains from 83 to 87% of carbon, from 10 to 16% of hydrogen, and small percentages of oxygen, nitrogen, and sulphur.   Some crude oils are devoid of sulphur and nitrogen, but all those obtained along the Pacific coast contain oxygen, sulphur, nitrogen, and a small percentage of moisture. The presence of sulphur in an oil is manifested by a very disagreeable odor. The following analyses of crude oils from Beaumont, Texas, and Bakersfield, Cal., will serve to give an idea as to the composition of the oils from these fields.

Constituents of Crude Oil	Texas Per Cent.	California Per Cent.
Carbon	84.60	85.0
Hydrogen	10.90	12.0
Sulphur	1.63	8
Oxygen	2.87	1.0
Nitrogen		2
Moisture		1.0

Properties of Fuel Oil.—Owing to the great demand for gasoline in all its various grades, the better grades of petroleum, such as those obtained in Pennsylvania, are treated so as to recover the lighter hydrocarbons. Of these, gasoline is among the early distillates, and when the gasoline, naphtha, and kerosene have been separated, the residue contains the lubricating oils, paraffin, and coke. This residue may be further distilled, so as to obtain the several products named; or, it may be used as a fuel, being then termed *fuel oil*. In other words, fuel oil is simply the heavier compounds of carbon and hydrogen contained in crude petroleum, the lighter compounds having been driven off by distillation. The analysis of a fuel oil derived from Beaumont crude oil is as follows:

Constituents	*Per Cent.*
Carbon	83.26
Hydrogen	12.41
Sulphur	50
Oxygen	3.83

On comparing this analysis with that of Beaumont crude oil previously given, it will be seen that the relative proportions of hydrogen and carbon have been changed and that a large part of the sulphur has been eliminated by the treatment of the crude oil.

Calorific Values of Oil Fuels.—The combustible elements contained in oil fuels are the same as those in coal, namely, carbon and hydrogen, and possibly a small proportion of sulphur. The heat of combustion per pound of oil, or the calorific value, may be found approximately from the chemical analysis of the oil by the same formula as that used for finding the heat value of coal. A more accurate method, however, is to burn a known weight of oil in a calorimeter and to measure the heat generated, from which the heat per pound of the oil may readily be calculated. From the results of available tests it is found that the heat of combustion per pound of oil fuel varies between 17,000 and 21,000 B. T. U. The average calorific value of Texas and California crude oils seems to be about 18,600 B. T. U. per lb.

Atomization of Oil.—When coal is used under steam boilers, the furnace contains a considerable amount of fuel; but early experiments with liquid fuel soon proved that the methods adopted for solid fuel were not applicable to liquid fuel, and that the latter could not be burned successfully in bulk. To insure satisfactory burning of oil fuel, it must first be changed to a vapor, and this is now accomplished by atomizing the oil, or converting it into the form of a very minutely divided spray. It is the vapor that burns, and not the liquid oil itself. If a sliver of burning wood is thrust into an open pan of fuel oil, the oil will not ignite, and the flame of the stick will be extinguished. The reason is that insufficient oil surface is exposed to the action of heat, and vaporization does not occur rapidly enough to supply the necessary quantity of inflammable gases to support combustion. By atomizing the oil, each minute particle is exposed to the air, thus providing for rapid evaporation and complete combustion.

Mixture of Oil Spray and Air.—Having changed the oil to a spray or to a vapor, it is next necessary to mix it intimately with air in the correct ratio to produce complete combustion. There are different methods by which the air is admitted so as to accomplish the mixing. Sometimes it is allowed to enter through holes surrounding the spraying devices and sometimes through openings from the ash-pit into the furnace; combinations of both methods may also be used. In any event, the main object to be attained is the thorough mixing of the spray and the air, so that each particle of oil shall be surrounded by the oxygen required for its perfect combustion.

Comparison of Steam and Air for Atomizing.—In stationary-boiler practice the agent most extensively used for the atomization of oil fuel is steam, air being used in rare or special cases. There seems to be little, if any, saving of fuel by using compressed air for atomizing, for the reason that it requires about the same amount of steam to operate the air compressor as to atomize the oil directly. Moreover, with the direct use of steam there is less complication of apparatus than when a compressor is installed, and there is a correspondingly smaller risk of accidents that may interrupt the service. Also the installation required for atomization by air is considerably more expensive than that required for the application of steam.

Amount of Steam Used for Atomizing.—During 1902 and 1903 the Bureau of Steam Engineering of the U. S. Navy Department made an extensive series of tests of oil burners using steam and air as atomizing agents. From the results of these tests it is found that the atomization of each pound of oil required the use of from .15 to 1 lb. of steam, the average value being about .55 lb. For good performance, the value should lie between .3 and .5 lb. The amount of steam used for atomization, expressed as a percentage of the total amount of steam generated by the boiler, ranged from 1 to 10%, but the average was approximately 2%. The foregoing values refer to the steam used for atomization only, whether directly or through the medium of the compressor, and are representative of average practice. They do not include the steam required for the oil-pressure pumps, which amounts to another 2%, approximately. Hence, in calculating capacities for an installation, it will be safe to assume that 5% of the steam generated will be utilized by the atomizing and pressure systems.

Effect of Steam on Combustion.—The steam that is used for the purpose of atomizing the oil does not increase the total heat resulting from combustion, although it may affect the character of the chemical changes in certain parts of the flame so as to produce a higher temperature at those points. An impression that seems to have gained credence is that the steam, under the effects of the high furnace temperature, is dissociated into its elements, oxygen and hydrogen, and that the combustion of the hydrogen thus set free increases the heat of combustion. The impression is wrong, for it takes just as much heat to break the steam up into its elements as is obtained by the subsequent uniting of those elements. Consequently, if the combustion is perfect, the steam that enters the furnace passes up the stack as steam, carrying away heat with it, and the greater the amount of steam introduced, the greater will be the heat loss. Thus, the introduction of steam into the furnace decreases the available heat rather than increases it.

Excess of Air in Oil Burning.—A pound of oil fuel of average composition requires about 13 or 14 lb. of air for its complete combustion; however, a greater amount must be admitted to the furnace to insure complete combustion, as the mixture of the air and the oil is not perfect.   The excess of air is required in order that the combustible elements may be surrounded by sufficient oxygen during the subsequent mixing in the combustion chamber.   The percentage of this excess should be as small as can be obtained without forming smoke.   In some boiler tests the excess has amounted to only 10%, which is extremely low; but under average conditions the excess of air is usually over 15%.   When coal is used as fuel, the excess of air is from 50 to 100%, or more; therefore, it is not surprising that a fireman accustomed to burning coal should admit too much air when burning oil fuel.   To serve as a guide to the fireman in the regulation of combustion, it is a good plan to install a CO_2 recorder, which will give a continuous record of the percentage of CO_2 in the flue gases.   The amount of CO_2 formed bears a known relation to the amount of air admitted, and by instructing the fireman to obtain as high a percentage of CO_2 as possible, the economy of operation may be increased. In practice, about 15% of CO_2 indicates the best performance obtainable, and an average of from 12 to 13% may be considered very satisfactory.

Evaporative Power of Oil Fuel.—Owing to its higher calorific value, oil fuel has a greater evaporative capacity, per lb., than is possessed by coal.   Moreover, the conditions under which oil fuel is burned enable a greater proportion of the heating value to be obtained for evaporation of the water than is the case with coal.   As a consequence, the number of pounds of water evaporated per lb. of oil is greater than the evaporation per lb. of coal.   Tests of boilers using a good grade of coal have shown an evaporation of slightly more than 11 lb. of water per lb. of coal and, under particularly favorable conditions, even better results have been obtained.   On the other hand, an average evaporation of from 12 to 13 lb. of water per lb. of oil has frequently been obtained with oil fuel, and in some cases an evaporation of 16 lb. has been reached.

Effect of Sulphur in Oil.—The presence of sulphur in oil fuels, particularly in crude oils, has caused some engineers to fear that the plates and tubes of oil-burning boilers would be pitted and corroded by the sulphurous gases generated.   Their fears, however, seem to have been groundless, inasmuch as the boiler inspection and insurance companies, who would be most likely to know, have made mention of no cases of excessive pitting or corrosion directly traceable to the presence of sulphur in oil fuel. Some of the lower grades of bituminous coal, containing from 2 to 4% of sulphur, have been used in steam-boiler furnaces without detriment to the boilers, although the grate bars may have been affected. Consequently, there seems to be no good reason why oil containing sulphur cannot be burned with the same freedom from deteriorating effects on the boiler.

Flash Point and Firing Point.—If a sample of fuel oil or of crude oil is placed in an open cup and heat is applied, the oil will begin to vaporize and inflammable gases will be driven off. If, while the heating proceeds, a lighted match is passed at intervals over the surface of the oil and about $\frac{1}{2}$ in. from it, a point will be reached at which the vapor rising from the oil will ignite and burn with a flicker of blue flame. The temperature of the oil when this flame first becomes apparent is termed the *flash point* of the oil. If the heating of the sample is continued, the vapors will be given off more rapidly and eventually they will ignite and burn continuously at the surface of the oil when the lighted match is brought near. The temperature of the oil when the burning becomes continuous is termed the *firing point* of the oil. The flash point and the firing point of an oil depend on the composition, specific gravity, and source of the oil.

Specifications for Oil Fuel.—An oil to be used as a fuel for steam boilers may be either a crude oil of uniform composition or a fuel oil. If it is the latter, all constituents having a low flash point will have been removed by distillation. The distillation should not have been carried on at a temperature high enough to burn the oil or to cause particles of carbon to separate from the oil; for these carbon particles will eventually clog the pipes and burners and cause trouble. The flash point of an oil fuel, as determined by a standard testing apparatus, should not be below 140° F.; otherwise, there will be danger from the inflammable vapors given off. The percentage of water in the oil should be less than 2, and the percentage of sulphur should not exceed about 1. If these proportions are exceeded, the oil will be more troublesome to use and should be purchasable at a correspondingly lower price.

Firebrick Lining of Furnaces.—It is possible to burn oil fuel by spraying the oil directly into the metallic firebox of an internally fired boiler or the ordinary furnace of an externally fired boiler; but, although it may be done and occasionally is done, it is not good practice and is not recommended. The spray of oil issuing from the burner should not strike the tubes or the comparatively cold metal surfaces of the boiler, but should first be completely burned in a combustion chamber of ample size, after which the hot gases may be led into contact with the heating surface of the boiler. A carefully designed furnace is either partly or wholly lined with firebrick, which protects the boiler from the direct action of the flames, prevents the hot gases from being chilled before combustion is complete, and tends to produce a more uniform transmission of heat to the boiler.

Effect of Firebrick Lining on Combustion.—Under the effect of the high temperature of combustion of oil fuel, the firebrick lining of the furnace is maintained in an incandescent state, which is of advantage in that it tends to promote a more nearly uniform flow of heat to the boiler. If there is dirt or water in the oil supply, or if the oil pumps do not act properly, so that the oil supply is variable in pressure or not continuous, the burners will act in a gusty, erratic manner. Under such circumstances, with an unlined firebox or furnace, it would be difficult and troublesome, if not impossible, to maintain combustion; but with a lining of incandescent firebrick there is a reserve of heat in the furnace, so that combustion will be restarted in case the fuel supply is momentarily interrupted by dirt or water. Moreover, the firebrick, acting as a heat reservoir, makes the flow of heat to the boiler more regular, without seriously reducing the heat-transmitting efficiency of the plates or tubes that it covers.

Furnace Proportions for Oil Fuel.—It is not necessary to observe any definite ratio of length to breadth or length to depth of the combustion space. The important point is to provide ample volume, and then to insure that the gases fill it in all parts and have the same velocity of flow throughout it.

STEAM

. ———

PROPERTIES OF STEAM

Saturated Steam.—If water is put in a closed vessel and heat is applied until boiling occurs and steam is given off, the pressure and the temperature of the steam will be the same as those of the water. The steam thus produced is known as *saturated steam;* that is, saturated steam is steam whose temperature is the same as that of boiling water subjected to the same pressure. Its nature is such that any loss of heat will cause some of the steam to condense, provided the pressure is not changed. Saturated steam that carries no water particles with it is called *dry saturated steam;* if it contains moisture, it is called *wet steam.* At every different pressure, saturated steam has certain definite values for the temperature, the weight per cubic ft., the heat per lb., and so on. These various values, collected and arranged in order, form the table of the Properties of Saturated Steam, more commonly termed the Steam Table. This table is shown on the following pages.

The various properties of steam, with their symbols, as given in the Steam Table, are as follows:

1. The temperature, t, of the steam, which is the boiling point of the water from which the steam is formed.

2. The *heat of the liquid*, q, which is the number of B. T. U. required to raise the temperature of 1 lb. of water from 32° F. to the boiling point corresponding to the given pressure.

3. The *latent heat of vaporization*, r, often termed the *latent heat*, which is the number of B. T. U. required to change 1 lb. of water at the boiling point into steam at the same temperature.

4. The *total heat of vaporization*, H, often termed the *total heat*, which is the number of B. T. U. required to raise 1 lb. of water from 32° F. to the boiling point for any given pressure and to change it into steam at that pressure. It is the sum of the heat of the liquid and the latent heat.

PROPERTIES OF SATURATED STEAM

| Absolute Press. Lb. per Sq. In. | Fahrenheit Temperature | British Thermal Units | | | Volume of 1 lb. in Cu. Ft. | Weight of 1 Cu. Ft. in Pounds |
| | | Heat of the Liquid from 32° F. | Total Heat from 32° F. | Latent Heat of Vaporization | | |
p	.	q	H	r	V	w
1	101.99	70.0	1,113.1	1,043.0	334.6	.00299
2	126.27	94.4	1,120.5	1,026.1	173.6	.00576
3	141.62	109.8	1,125.1	1,015.3	118.4	.00844
4	153.09	121.4	1,128.6	1,007.2	90.31	.01107
5	162.34	130.7	1,131.5	1,000.8	73.22	.01366
6	170.14	138.6	1,133.8	995.2	61.67	.01622
7	176.90	145.4	1,135.9	990.5	53.37	.01874
8	182.92	151.5	1,137.7	986.2	47.07	.02125
9	188.33	156.9	1,139.4	982.5	42.13	.02374
10	193.25	161.9	1,140.9	979.0	38.16	.02621
11	197.78	166.5	1,142.3	975.8	34.88	.02866
12	201.98	170.7	1,143.6	972.9	32.14	.03111
13	205.89	174.6	1,144.7	970.1	29.82	.03355
14	209.57	178.3	1.145.8	967.5	27.79	.03600
14.7	212.0	180.8	1,146.6	965.8	26.60	.03760
16	216.32	185.1	1,147.9	962.8	24.59	.04067
18	222.40	191.3	1,149.8	958.5	22.00	.04547
20	227.95	196.9	1,151.5	954.6	19.91	.05023
22	233.06	202.0	1,153.0	951.0	18.20	.05495
24	237.79	206.8	1,154.4	947.6	16.76	.05966
26	242.21	211.2	1,155.8	944.6	15.55	.06432
28	246.36	215.4	1,157.1	941.7	14.49	.06899
30	250.27	219.4	1,158.3	938.9	13.59	.07360
32	253.98	223.1	1,159.4	936.3	12.78	.07820
34	257.50	226.7	1,160.4	933.7	12.07	.08280
36	260.85	230.0	1,161.5	931.5	11.45	.08736
38	264.06	233.3	1,162.5	929.2	10.88	.09191
40	267.13	236.4	1,163.4	927.0	10.37	.09644
42	270.08	239.3	1,164.3	925.0	9.906	.1009
44	272.91	242.2	1,165.2	923.0	9.484	.1054
46	275.65	245.0	1,166.0	921.0	9.907	.1099
48	278.30	247.6	1,166.8	919.2	8.740	.1144
50	280.85	250.2	1,167.6	917.4	8.414	.1188
52	283.32	252.7	1.168.4	915.7	8.110	.1233
54	285.72	255.1	1,169.1	914.0	7.829	.1277
56	288.05	257.5	1,169.8	912.3	7.568	.1321
58	290.31	259.7	1,170.5	910.8	7.323	.1366
60	292.51	261.9	1,171.2	909.3	7.096	.1409
62	294.65	264.1	1,171.8	907.7	6.882	.1453
64	296.74	266.2	1,172.4	906.2	6.680	.1497

TABLE—(*Continued*)

p	.	q	H	ı	V	w
66	298.78	268.3	1,173.0	904.7	6.490	.1541
68	300.76	270.3	1,173.6	903.3	6.314	.1584
70	302.71	272.2	1,174.3	902.1	6.144	.1628
72	304.61	274.1	1,174.9	900.8	5.984	.1671
74	306.46	276.0	1,175.4	899.4	5.834	.1714
76	308.28	277.8	1,176.0	898.2	5.691	.1757
78	310.06	279.6	1,176.5	896.9	5.554	.1801
80	311.80	281.4	1,177.0	895.6	5.425	.1843
82	313.51	283.2	1,177.6	894.4	5.301	.1886
85	316.02	285.8	1,178.3	892.5	5.125	.1951
90	320.04	290.0	1,179.6	889.6	4.858	.2058
95	323.89	294.0	1,180.7	886.7	4.619	.2165
100	327.58	297.9	1,1 1.9	884.0	4.403	.2271
105	331.13	301.6	1,182.9	881.3	4.206	.2378
110	334.56	305.2	1,184.0	878.8	4.026	.2484
115	337.86	308.7	1,185.0	876.3	3.862	.2589
120	341.05	312.0	1,186.0	874.0	3.711	.2695
125	344.13	315.2	1,186.9	871.7	3.572	.2800
130	347.12	318.4	1,187.8	869.4	3.444	.2904
135	350.03	321.4	1,188.7	867.3	3.323	.3009
140	352.85	324.4	1,189.5	865.1	3.212	.3113
145	355.59	327.2	1,190.4	863.2	3.107	.3218
150	358.26	330.0	1,191.2	861.2	3.011	.3321
155	360.86	332.7	1,192.0	859.3	2.919	.3426
160	363.40	335.4	1,192.8	857.4	2.833	.3530
165	365.88	338.0	1,193.6	855.6	2.751	.3635
170	368.29	340.5	1,194.3	853.8	2.676	.3737
175	370.65	343.0	1,195.0	852.0	2.603	.3841
180	372.97	345.4	1,195,7	850.3	2.535	.3945
185	375.23	347.8	1,196.4	848.6	2.470	.4049
190	377.44	350.1	1,197.1	847.0	2.408	.4153
195	379.61	352.4	1,197.7	845.3	2.349	.4257
200	381.73	354.6	1,198.4	843.8	2.294	.4359
205	383.82	356.8	1,199.0	842.2	2.241	.4461
210	385.87	358.9	1,199.6	840.7	2.190	.4565
215	387.88	361.0	1,200.2	839.2	2.142	.4669
220	389.84	363.0	1,200.8	837.8	2.096	.4772
225	391.79	365.1	1,201.4	836.3	2.051	.4876
230	393.69	367.1	1,202.0	834.9	2.009	.4979
235	395.56	369.0	1,202.6	833.6	1.968	.5082
240	397.41	371.0	1,203.2	832.2	1.928	.5186
250	400.99	374.7	1,204.2	829.5	1.854	.5393
260	404.47	378.4	1,205.3	826.9	1.785	.5601
275	409.50	383.6	1,206.8	823.2	1.691	.5913
300	417.42	391.9	1,209.3	817.4	1.554	.644
325	424.82	399.6	1,211.5	811.9	1,437	.696

5.   The *specific volume*, *V*, which is the volume, in cubic ft., of 1 lb. of steam at the given pressure.

6.   The *density*, *w*, which is the weight, in pounds, of 1 cu. ft. of steam at the given pressure.   It is the reciprocal of the specific volume.

The pressures, *p*, given in the first column of the Steam Table, are absolute pressures.   The pressure registered by the gauge on the boiler is the *gauge pressure*, or the pressure of the steam above that of the atmosphere.   The pressure of the atmosphere at sea level, with the barometer at about 30 in., is approximately 14.7 lb. per sq. in.   Therefore, the *absolute pressure* at sea level is equal to the gauge pressure plus 14.7.   In using the Steam Table, the atmospheric pressure, 14.7 lb. per sq. in., must always be added to the gauge pressure.

Use of Steam Table.—For any absolute pressure *p* given in the first column of the Steam Table, the corresponding temperature *t*, total heat *H*, or other property is found in the *same* horizontal line, under the proper column heading; but if the pressure lies between two of the values given in the first column, the corresponding temperature, total heat, etc. must be found by the process known as *interpolation*, as illustrated in the following examples:

EXAMPLE 1.—Find the temperature corresponding to a pressure of 147 lb. per sq. in., absolute.

SOLUTION.—Referring to the Steam Table,

for $p = 150$ lb., $t = 358.26°$

and   for $p = 145$ lb., $t = 355.59°$

Difference, 5 lb.,   2.67°

Difference for 1 lb. difference of pressure is $\dfrac{2.67°}{5} = 534°.$

147 lb. − 145 lb. = 2 lb., the given difference of pressure; and for this, the difference in temperature is $2 \times .534° = 1.068°$ or 1.07°, taking two decimal places.   Hence, the increase of 2 lb. from 145 lb. to 147 lb. is accompanied by an increase in temperature of 1.07°.   Therefore, adding the increase 1.07° to the temperature 355.59° corresponding to 145 lb., the temperature for 147 lb. is $355.59° + 1.07° = 356.66°.$

EXAMPLE 2.—The pressure in a steam boiler as shown by the gauge is 87 lb. per sq. in. What is the temperature of the steam?

SOLUTION.—The absolute pressure is $87+14.7=101.7$ lb. per sq. in. This pressure, in the Steam Table, lies between the values 100 and 105.

$$\text{for } p = 105 \text{ lb.}, \; t = 331.13°$$
$$\text{for } p = 100 \text{ lb.}, \; t = 327.58°$$
$$\overline{\text{Difference, 5 lb.,} \quad\; 3.55°}$$

For 1 lb. change of pressure, the difference in temperature is $\dfrac{3.55°}{5}=.71°$. From 100 lb. to 101.7 lb., the change of pressure is 1.7 lb., and the corresponding change of temperature is $.71°\times1.7=1.207°$, or $1.21°$ as the values in the Steam Table contain but two decimal places. For 101.7 lb., therefore, the temperature is $327.58°+1.21°=328.79°$.

EXAMPLE 3.—What is the pressure of steam at a temperature of 285° F.?

SOLUTION.—From the Steam Table,

$$\text{for } t = 285.72°, \; p = 54 \text{ lb.}$$
$$\text{for } t = 283.32°, \; p = 52 \text{ lb.}$$
$$\overline{\text{Difference, 2.40°,} \quad 2 \text{ lb.}}$$

From $t=283.32°$ to $t=285°$, the increase of temperature is $1.68°$. Now, since an increase of temperature of $2.40°$ gives an increase of pressure of 2 lb., the increase of $1.68°$ must give an increase of pressure of $\dfrac{1.68}{2.40}\times2$ lb. $=1.4$ lb. Hence, the required pressure is 52 lb. $+1.4$ lb. $=53.4$ lb.

EXAMPLE 4.—Find, from the Steam Table, the total heat of a pound of saturated steam at a pressure of 63 lb. per sq. in., gauge.

SOLUTION.—The absolute pressure is $\bar{63}+14.7=77.7$ lb. per sq. in. From the Steam Table,

$$\text{for } p = 78 \text{ lb.}, \; H = 1{,}176.5 \text{ B. T. U.}$$
$$\text{for } p = 76 \text{ lb.}, \; H = 1{,}176.0 \text{ B. T. U.}$$

$$\overline{\text{Difference, 2 lb.,} \qquad .5 \;\; \text{B. T. U.}}$$
$$\text{Difference, 1 lb.,} \qquad .25 \text{ B. T. U.}$$

8

The difference between the given pressure and 76 lb. is 77.7 −76 = 1.7 lb. For a difference of 1.7 lb., the change of total heat is 1.7×.25 = .425 B. T. U. Hence, for 77.7 lb., H = 1,176.0 +.425 = 1,176.425, say 1,176.4 B. T. U

EXAMPLE 5.—Find the volume occupied by 14 lb. of steam at 30 lb. gauge pressure.

SOLUTION.—Absolute pressure = 30+14.7 = 44.7 lb. per sq. in. From the Steam Table,

for p = 44 lb., V = 9.484 cu. ft.
for p = 46 lb., V = 9.097 cu. ft.

Difference, 2 lb.,   .387 cu. ft.

The difference for 1 lb. is $\dfrac{.387}{2}$ = .1935.   44.7−44 = .7 lb.

actual difference in pressure. .1935×.7 = .135 difference in volume. As the pressure increases, the volume decreases; and to obtain the volume at 44.7 lb., it is necessary to subtract the difference .135 from the volume at 44 lb.; thus, for p = 44.7, V = 9.484−.135 = 9.349 cu. ft. The volume of 14 lb. is 14×9.349 cu. ft. = 130.89 cu ft.

EXAMPLE 6.—Find the weight of 40 cu. ft. of steam at a temperature of 254° F.

SOLUTION.—From column 10 of the Steam Table, the weight w of 1 cu. ft. of steam at 253.98 is .07820 lb. 254−253.98 = .02. Neglecting the .02°, the weight of 40 cu. ft. is therefore .07820×40 = 3.128 lb.

EXAMPLE 7.—How many pounds of steam at 64 lb. pressure, absolute, are required to raise the temperature of 300 lb. of water from 40° to 130° F., the water and steam being mixed together?

SOLUTION.—The number of heat units required to raise 1 lb. from 40° to 130° is 130°−40° = 90 B. T. U. Actually, a little more than 90 would be required but the above is near enough for all practical purposes. Then, to raise 300 lb. from 40° to 130° requires 90×300 = 27,000 B. T. U. This quantity of heat must necessarily come from the steam. Now, 1 lb. of steam at 64 lb. pressure gives up, in condensing, its latent heat of vaporization, or 906.2 B. T. U.; but, in addition to its latent heat, each pound of steam on condensing must give up an additional amount of heat in falling to 130°. Since the original temperature of the steam was 296.74° F. (see Steam Table), each pound gives up by its fall of temperature 296.74 − 130 = 166.74 B. T. U. Consequently, each pound of the steam gives up a total of 906.2+166.74 = 1,072.94 B. T. U., and $\frac{27,000}{1,072.94} = 25.16$ lb. of steam will therefore be required to accomplish the desired result.

SUPERHEATED STEAM

If saturated steam is contained in a vessel, out of contact with water, and heat is added to it, its temperature will begin to rise and its weight per cu. ft. will begin to decrease, provided the pressure remains constant. As more heat is added, the temperature rises farther above that of saturated steam at that pressure, and the steam is then called *superheated steam*. Superheated steam cannot exist in contact with water.

The following distinction is usually made between saturated and superheated steam: For a given pressure, saturated steam has one temperature and one weight per cu. ft., neither of which can change so long as the steam remains in immediate contact with water. Superheated steam at the same pressure has a greater temperature and less weight per cu. ft. than saturated steam, and both the temperature and weight per cu. ft. may vary while the pressure remains constant if the volume increases or decreases accordingly. In other words, both the pressure and the volume of superheated steam must be constant in order to maintain a constant temperature and a constant weight per cu. ft.

QUALITY OF STEAM

Moisture in Steam.—The steam furnished by the average steam boiler is not dry saturated steam, but is usually wet steam. A good boiler should not show more than 2 or 3% of water in the steam. In a quantity of wet steam, or a mixture of steam and water, the percentage of dry steam, expressed as a decimal, is called the *quality of the steam*. For example, suppose that a certain boiler generates wet steam that contains 3%, or .03, of moisture; then the quality of the steam, or the percentage of dry steam, is .97. In other words, the quality of the steam is equal to 1 minus the percentage of moisture, expressed decimally. This rule may be stated simply by the formula

$$Q = 1 - m,$$

in which Q = quality of the steam;

m = percentage of moisture, expressed decimally.

EXAMPLE.—What is the quality of steam that contains 2.7% of moisture?

SOLUTION.—Expressed as a decimal, 2.7% = .027. Then, substituting this value for m in the formula, $Q = 1 - .027 = .973$.

Heat in Wet Steam.—The total heat contained in 1 lb. of dry steam is the sum of the heat required to raise 1 lb. of water from 32° F. to the boiling point and the heat required to change the boiling water into steam of the same temperature. That is, in the Steam Table, each value given in the fourth column is the sum of the values given in the third and fifth columns and lying in the same horizontal row. In a mixture of 1 lb. of steam and water at the same temperature there is less heat than in 1 lb. of dry steam at the same temperature; for all the water has not been changed to steam, and consequently the latent heat of 1 lb. of steam has not been utilized. Instead, there is present only that part of the latent heat which is used to evaporate the portion of the mixture that is dry steam, which is represented by the quality of the steam. Thus, using the symbols given in the Steam Table,

$$H = q + r \qquad (1)$$

which is the formula for the total heat of 1 lb. of dry steam. But if the steam is wet, and Q represents the quality of the steam, expressed decimally, the total heat of 1 lb., represented by H^1, is $\qquad H^1 = q + Qr \qquad (2)$

EXAMPLE.—What is the total heat of 10 lb. of steam at 150 lb. gauge pressure, if the steam contains 5% of moisture?

SOLUTION.—From the Steam Table, the heat of the liquid of 1 lb. of dry steam at 150 lb., gauge, or $150 + 14.7 = 164.7$ lb., absolute, is $q = 337.84$ B. T. U., and the latent heat of 1 lb. at the same pressure is $r = 855.71$ B. T. U. As the moisture is 5%, the quality of the steam is $1.00 - .05 = .95$. Then, applying formula 2, $H^1 = 337.84 + .95 \times 855.71 = 1,150.76$ B. T. U.

Barrel Calorimeter.—It is a rather difficult matter to make a very exact determination of the moisture contained in steam. The apparatus or instrument used for this purpose is called a *calorimeter.* There are many more or less complicated calorimeters, but about the simplest and most available one for general use is the so-called *barrel calorimeter,* shown in the accompanying illustration. A barrel or tank *a* holding 400 or 500 lb. of water is placed on a platform scales *b*, filled with water, and weighed. The temperature of the water is registered by a thermometer inserted in the side of the barrel. Steam from the boiler is led through a hose *c* into the barrel until the temperature of the water

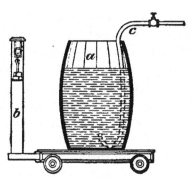

reaches 130° to 140° F. The steam is then turned off and the barrel and its contents are again weighed. The difference between this weight and the original weight is the weight of the steam led in from the boiler and condensed in the barrel. The average steam pressure throughout the process must be observed. It is well to have the tube bent as shown in the figure.

The weight of the cold water and the rise in its temperature are known, and so also is the weight of the mixture of steam and water that is led in from the boiler. From the Steam Table, the temperature of the steam can be found, since the average pressure is known. Now, if dry steam comes through the hose *c*, the condensation of this steam should raise the temperature of the cold water in the barrel a certain amount. If the temperature is not raised that much, it must be because some of the mixture led into the barrel was water.

Let Q = quality of steam;

W = weight of cold water in barrel, in pounds;

w = weight of mixture run into the barrel, in pounds;

t = temperature of steam corresponding to the observed pressure;

t_1 = original temperature of water, in barrel;

t_2 = temperature of water in barrel after steam is condensed;

L = latent heat of 1 lb. of steam at observed pressure.

Then, $$Q = \frac{\dfrac{W\,(t_2 - t_1)}{w} - (t - t_2)}{L}$$

EXAMPLE.—In a calorimetric test, the weight of cold water
was 420 lb., and of steam condensed, 36 lb. The initial
temperature of the cold water was 40° F., the final temperature was 130° F., and the average steam pressure was 60 lb.
Find the quality of the steam.

SOLUTION.—Absolute pressure = 60 + 14.7 = 74.7 lb. per sq. in.
Latent heat of steam at this pressure, from Steam Table, is
898.5. The temperature of steam at this pressure is 307.2°
Hence, by the formula

$$Q = \frac{\dfrac{420\,(130-40)}{36} - (307.2 - 130)}{898.5} = .9714$$

That is, the boiler generates a mixture that is composed of
97.14% of dry steam and 2.86% of water.

If the result found by the foregoing formula shows that Q
is greater than 1, the conclusion is that the steam, instead of
being wet, was superheated, and therefore gave up, in condensing, a greater amount of heat per pound than would have
been given up by 1 lb. of dry saturated steam at the same
pressure.

The barrel calorimeter must be used very carefully in order
to obtain accurate results. The operation should be repeated
once or twice before the actual test is made so as to warm up
the barrel. The most important observation is the temperature. This should be taken by a thermometer graduated to
fifths or tenths of a degree. The weights should be as accurate as possible. The chief merit of the barrel calorimeter is
its availability. It can be rigged up in almost any situation.

The quality of the steam having been determined, the actual
amount of water evaporated by a steam boiler is found by
multiplying the observed amount of feedwater by the quality
of the steam expressed decimally.

FLOW OF STEAM

Weight of Steam Discharged.—The number of pounds of steam that will flow continuously through a pipe of given diameter in 1 min. at specified pressure may be calculated by the formula

$$W = 87 \sqrt{\frac{w(P_1 - P_2)\ d^5}{L\left(1 + \dfrac{3.6)}{d}\right)}}$$

in which W = weight of steam discharged, in pounds per minute;
  w = weight of 1 cu. ft. of steam at the pressure P_1;
  P_1 = pressure of steam at entrance to pipe, in pounds per square inch;
  P_2 = pressure of steam at discharge, in pounds per square inch;
  L = length of pipe, in feet;
  d = diameter of pipe, in inches.

In applying the preceding formula in determining the diameter of the steam pipe for an engine, it must be remembered that, in steam-engine work, the steam is drawn intermittently from the pipe. Thus, assume that an engine of 100 H. P., consuming 30 lb. of steam per horsepower per hour, cuts off at ¼ stroke. In that case, the steam consumption per hour would be $100 \times 30 = 3,000$ lb. But as the steam used at each stroke is drawn into the cylinder during only one-fourth of the time required to complete the stroke, the 3,000 lb. of steam flows through the pipe in ¼ hr. Then, in order to determine the quantity of steam that would flow continuously at the same velocity at which it flows during admission to the cylinder, the actual steam consumption per hour should be divided by the fraction representing the cut-off and the quotient should be taken as the weight of steam discharged per hour. This value, divided by 60, should be substituted for w in the formula. Thus, in the case mentioned, the amount of steam discharged per hour, flowing continuously at the same velocity as during the admission period, is $3,000 \div \frac{1}{4} = 12,000$, and the value of W to be used in the formula is therefore $12,000 \div 60 = 200$ lb. per min. Knowing the pressures, the length of pipe, and the weight of the entering steam per cubic foot, different values of d may be assumed, until a value is found that will give the necessary discharge W. This is the required pipe diameter.

The approximate weights of steam delivered per minute through 100 ft. of pipe of various diameters, with a drop of pressure of 1 lb., are given in the accompanying table. On the whole, these values are slightly higher than those which would be obtained by the foregoing formula for the same conditions. If the drop of pressure is more or less than 1 lb., the value in the table must be multiplied by the square root of the drop, to obtain the discharge. Also, if the length of the pipe is more or less than 100 ft., divide 100 by the length, in feet, and multiply the square root of this quotient by the value given in the table. The following example illustrates this point.

EXAMPLE.—How many pounds of steam will be discharged per minute, with an initial gauge pressure of 120 lb. per sq. in., through a pipe 3 in. in diameter and 400 ft. long, with a drop of pressure of 2 lb.?

SOLUTION.—From the table, the amount discharged through 100 ft. of 3-in. pipe with a drop of 1 lb. and an initial pressure of 120 lb. per sq. in., is 53.6 lb. per min. But as the drop is 2 lb., the table value must be multiplied by $\sqrt{2}$ and as the length is 400 ft., it must also be multiplied by $\sqrt{\tfrac{100}{400}}$. Hence, the discharge for the given conditions will be $53.6 \times \sqrt{2} \times \sqrt{\tfrac{100}{400}}$ $= 37.9$ lb. per min.

Resistance of Elbows and Valves.—The presence of elbows, bends, and valves in a steam pipe increases the resistance to the flow of steam and thus increases the drop of pressure between the inlet and outlet ends. It has been found that the resistance caused by an elbow or a sharp bend is approximately the same as the resistance of a length of pipe equal to 60 times the diameter, and that a stop-valve has a resistance equal to that of a length of pipe of 40 diameters. In using the foregoing formula for the weight of steam discharged, therefore, the value of L should be the *equivalent length* of pipe, taking into account the bends and valves. The method of doing this is illustrated by the following example:

EXAMPLE.—What is the equivalent length of 300 ft. of 3-in. pipe containing four elbows and six stop-valves?

WEIGHT OF STEAM DELIVERED PER ▮▮ THROUGH 100 FT. OF PIPE WITH 1 LB. DROP OF PRESSURE

Nominal Diameter of Pipe, in Inches

Weight, in Pounds, of Steam Delivered ▮r ▮te Through 100 Ft. of Pipe with 1 Lb. Drop of ▮re

Initial Gauge Pressure Lb. per Sq. In.	3	3½	4	4½	5	6	7	8	9	10
70	43.2	64.5	91.7	124.3	168.7	277.2	410.5	577.2	793.3	1,051.7
80	45.5	68.0	96.6	130.9	177.7	292.1	432.5	608.2	835.8	1,108.4
90	47.6	71.2	101.2	137.2	186.3	306.0	453.3	637.3	875.9	1,161.3
100	49.7	74.3	105.7	143.2	194.4	319.8	473.6	665.1	914.1	1,211.8
110	51.7	77.3	109.9	148.9	202.1	332.2	491.8	691.5	950.4	1,259.9
120	53.6	80.2	113.9	154.4	209.5	344.3	509.9	717.6	985.1	1,306.3
130	55.4	82.9	117.8	159.7	216.7	356.1	527.4	741.6	1,019.6	1,351.1
140	57.2	85.5	121.5	164.7	223.6	367.4	544.6	765.7	1,052.9	1,393.9
150	58.9	88.1	125.2	169.6	230.2	378.3	560.2	787.7	1,082.6	1,428.1

SOLUTION.--Each elbow has a resistance the same as that of 60 diameters of pipe, or $60\times3=180$ in. $=15$ ft., and four elbows have the resistance of $4\times15=60$ ft. of pipe. Each stop-valve has a resistance that is equivalent to adding $40\times3=120$ in. $=10$ ft. of pipe, and, as there are six valves, their combined resistance is that of $6\times10=60$ ft. of pipe. The equivalent length of pipe is, therefore, $300+60+60=420$ ft.

Steam Pipes for Engines.—In practice, the velocity of flow of steam in the supply pipes of engines and pumps is usually not greater than 6,000 ft. per min., although it is increased to as much as 8,000 ft. per min. in some cases. For exhaust pipes, a common value is 4,000 ft. per min. The assumptions made are that the cylinder is filled with steam at boiler pressure at each stroke and that a volume of steam equal to the volume of the cylinder is released at each stroke, so that the flow is practically continuous. The areas of the steam and exhaust pipes may then be calculated by the formula

$$a = \frac{A\,S}{s},$$

in which a = area of steam or exhaust pipe, in square inches;

A = area of cylinder, in square inches;

S = piston speed, in feet per minute;

s = velocity of steam in pipe, in feet per minute.

EXAMPLE.—Find the areas of the steam and exhaust pipes for an engine whose cylinder is 20 in. in diameter and whose piston speed is 450 ft. per min.

SOLUTION.—By the formula, the area of the steam pipe, assuming that $s=6,000$ ft. per min., is

$$a = \frac{.7854\times20^2\times450}{6,000} = 23.6 \text{ sq. in.}$$

Similarly, for the exhaust pipe, assuming that $s=4,000$, the area is

$$a = \frac{.7854\times20^2\times450}{4,000} = 35.3 \text{ sq. in.}$$

STEAM BOILERS

FURNACE FITTINGS

Bridge Wall.—The bridge, also termed the *bridge wall*, is a low wall at the back end of the grate; it forms the rear end of the furnace and causes the flame to come in close contact with the heating surface of the boiler. It is usually built of common brick and faced with firebrick. The passage between the bridge and the boiler shell should not be too small; its area may be approximately one-sixth the area of the grate. The space between the grate and the shell should be ample for complete combustion, and the distance between the grate and the boiler shell may be made about one-half the diameter of the shell.

Fixed Grates.—The grate, which is nearly always made of cast iron, furnishes a support for the fuel to be burned and must be provided with spaces for the admission of air. The area of the solid portion of the grate is usually made nearly equal to the combined area of the air spaces.

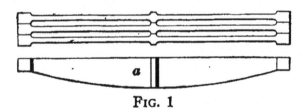

FIG. 1

The common type of fixed grate is made of single bars a, Fig. 1, placed side by side in the furnace. The thickness of the lugs cast on the sides of the bars determines the width of the open spaces of the grate. It is the general practice to make the thickness across the lugs twice the thickness of the top of the bar. For long furnaces, the bars are generally made in two lengths of about 3 ft. each, with a bearing bar in the middle of the grate. Long grates are generally set with a downward slope toward the bridge wall of about ¾ in. per foot of length.

For the larger sizes of anthracite and bituminous coal, the air space may be from ⅜ to ⅞ in. wide, and the grate bar may have the same width.   For pea and nut coal, the air space may be from ⅜ to ½ in., and for finely divided fuel, like buckwheat coal, rice coal, bird's-eye coal, culm, and slack, air spaces from 3/16 to ⅜ in. may be used.

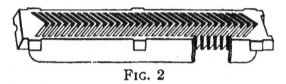

FIG. 2

The grate bar shown in Fig. 2, and known as the *herring-bone grate bar*, has in many places superseded the ordinary grate bar, because they will usually far outlast a set of ordinary grate bars.   Herring-bone grate bars can be obtained in a great variety of styles and with different widths of air spaces. They are usually supported on cross-bars, and, like many

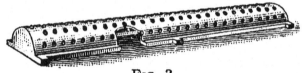

FIG. 3

other forms of grate bars, may be arranged with trunnions, so as to rock the individual bars by means of hand levers.

A form of cast-iron grate bar for the burning of sawdust is shown in Fig. 3.   The bar is semicircular in cross-section and is provided with circular openings for the introduction of air. Lugs are cast on each side of the bar to serve as distance pieces in providing air spaces between the bars.

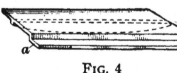

FIG. 4

Dead Plate.—The front ends of the grate bars are usually supported on the *dead plate*, which is a flat cast-iron plate placed across the furnace just inside the boiler front and on a level with the bottom of the furnace door.   The purpose of the dead plate is twofold: It forms a support for the firebrick lining of the boiler front, and a resting place on which bitumi-nous coal may be coked before it is placed on the fire. To support the grate bars, the inner edge of the dead plate is either beveled or a lip is provided, as at *a*, Fig. 4.

Objection to Stationary Grate Bars.—The greatest objec-tion to stationary grate bars is that with them the furnace door must be kept open for a considerable length of time when the fire is being cleaned. Cleaning fires when the boiler has a stationary grate not only severely taxes the fireman, but the inrush of cold air chills the boiler plates, thus producing stresses that in the course of time will crack them.

Shaking Grates.—There are on the market many designs of *shaking grates* for large steam boilers, differing chiefly in detail and arrangement. Usually the grate bars are hung on trun-nions at each end and are connected together with bars to which are attached shaking rods that extend forwards through the furnace front. Levers or handles are attached to the shaking rods, and by working them back and forth the grate bars receive a rocking motion that breaks up the bed of coal on the grate and serves to shake the ashes through into the ash-pit. The fires may thus be kept clean without the neces-sity of opening the fire-doors.

Classes of Mechanical Stokers.—A *mechanical stoker* is a power-driven rocking grate arranged so as to give a uniform feed of coal and to rid itself continuously of ashes and clinkers. The principal designs of mechanical stokers and automatic furnaces may be divided into two general classes, *overfeed* and *underfeed*.

Overfeed Stoker.—In overfeed stokers the fixed carbon of the coal is burned on inclined grates. The coal is pushed on to these grates, which are given a sufficiently rapid vibratory motion to feed it down at such a rate that practically all the carbon is burned before reaching the lower end, where the ashes and clinkers are discharged. In Fig. 5 is shown a sectional view of a stoker of this class. The coal is fed into the hopper *a*, from which it is pushed by the pusher plate *b* on to the dead plate *c*, where it is heated. From *c* it passes to the grate *d*. Each bar is supported at its ends by trunnions and is connected by an arm to a rocker bar *i*, which is slowly moved to and fro by an eccentric on the shaft *s*, so as to rock the grates back and forth; the grates thus gradually move the burning fuel downwards. The ashes and clinkers are discharged from the lower grate bar on to the dumping grate *e*. A guard *f* may be raised, as shown by the dotted lines, so as to prevent coke or coal from falling from the grate bars into the ash-pit when the dumping grate is lowered. Air for burning the gases is admitted in small jets through holes in the air tile *g*, and the mixture of gas and air is burned in the

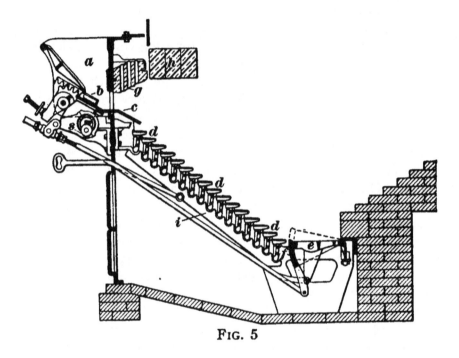

Fig. 5

hot chamber between the firebrick arch *h* and the bed of burning coke below.

Underfeed Stoker.—The stoker shown in Fig. 6 illustrates the principle of operation and the construction of the under-feed stoker. Coal is fed into the hopper *a*, from which it is drawn by the spiral conveyer *b* and forced into the magazine *d*. The incoming supply of fresh coal forces the fuel upwards to the surface and over the sides of the magazine on the grates, where it is burned. A blower forces air through a pipe *f* into the chamber *g* surrounding the magazine. From *g* the air passes upwards through hollow cast-iron tuyère blocks and out through the openings, or tuyères *e*. The gas formed in the magazine, mixed with the jets of air from the tuyères, rises through the burning fuel above, where it is subjected to a sufficiently high temperature to secure its combustion. Nearly all the air for burning the coal is supplied through the tuyères, only a very small portion of the supply coming through the grate. The

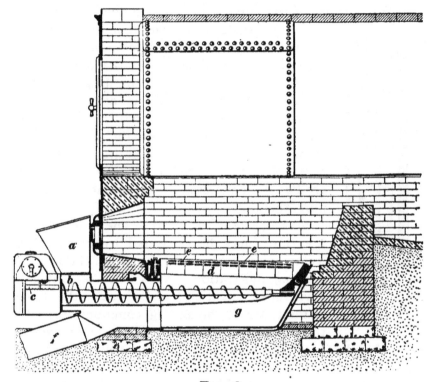

FIG. 6

ashes and clinkers are gradually forced to the sides of the grate against the side walls of the furnace, from which they are removed from time to time through doors in the furnace front similar to the fire-doors of an ordinary furnace. In other words, owing to the construction of the underfeed stoker, the fire must periodically be cleaned from clinkers and the ashes removed by hand.

CHIMNEYS

Production of Draft.—It is well known that any volume of gas is lighter when heated than the same volume of gas when cool. When hot gases pass into the chimney, they have a temperature of from 400° to 600° F., while the air outside the chimney has a temperature of from 40° to 90° F. Roughly speaking, the air weighs twice as much, bulk for bulk, as the hot gases. Naturally, then, the pressure in the chimney is less than the pressure of the outside air. The production of draft

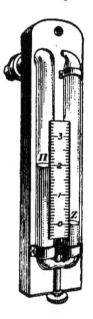

and the satisfactory operation of a chimney depend on this pressure difference. The pressure of the draft depends on the temperature of the furnace gases and the height of the chimney. Chimney draft is affected by so many varying conditions that no absolutely reliable rules can be given for proportioning chimneys to give a certain desired draft pressure. The rules given for chimney proportions are based on successful practice rather than on pure theory.

Measurement of Draft.—The intensity of the draft may be measured by means of a water gauge such as is shown in the accompanying illustration. As will be seen, it is a glass tube open at both ends, bent to the shape of the letter **U**; the left leg communicates with the chimney. The difference in the two water levels *H* and *Z* in the legs represents the intensity of the draft, and is expressed in inches of water.

The draft produced by a chimney may vary from ¼ in. to 2 in. of water, depending on the temperature of the chimney gases and on the height of the chimney. Generally speaking, it is advantageous to use a high chimney and as low a chimney temperature as possible. The draft pressure required depends on the kind of fuel used. Wood requires but little draft, say ⅛ in. of water or less; bituminous coal generally requires less draft than anthracite. To burn anthracite, slack, or culm, the draft pressure should be about 1¼ in. of water.

Form of Chimney.—The form of a chimney has a pronounced effect on its capacity. A round chimney has a greater capacity for a given area than a square one. If the flue is tapering, the area for calculation is measured at its smallest section. The flue through which the gases pass from the boilers to the chimney should have an area equal to, or a little larger than, the area of the chimney. Abrupt turns in the flue or contractions of its area should be carefully avoided. Where one chimney serves several boilers, the branch flue from each furnace to the main flue must be somewhat larger than its proportionate part of the area of the main flue.

Brick Chimneys.—Chimneys are usually built of brick, though concrete, iron, and steel are often used for those of moderate height. Brick chimneys are usually built with a flue having parallel sides and a taper on the outside of the chimney of from $\frac{1}{16}$ to $\frac{1}{4}$ in. per ft. of height. A round chimney gives greater draft area for the same amount of material in its structure and exposes less surface to the wind than a square chimney. Large brick stacks are usually made with an inner core and an outer shell, with a space between them. Such chimneys are usually constructed with a series of internal pilasters, or vertical ribs, to give rigidity. The top of the chimney should be protected by a coping of stone or a cast-iron plate to prevent the destruction of the bricks by the weather.

Iron and Steel Chimneys.—Iron or steel stacks are made of plates varying from $\frac{1}{8}$ to $\frac{1}{2}$ in. thick. The larger stacks are made in sections, the plates being about $\frac{1}{4}$ in. thick at the top and increasing to $\frac{1}{2}$ in. at the bottom; they are lined with firebrick about 18 in. thick at the bottom and 4 in. at the top. Sometimes no lining is used on account of the likelihood of corrosion and the difficulty of inspection, and also because the inside of lined stacks cannot be painted.

Chimney Foundations.—On account of the great concentration of weight, the foundation for a chimney should be carefully designed. Good natural earth will support from 2,000 to 4,000 lb. per sq. ft. The footing beneath the chimney foundation should be made of large area, in order to reduce the pressure due to the weight of the chimney and its foundation to a safe limit.

9

Height of **Chimney.**—The relation between the height of the chimney and the pressure of the draft, in inches of water, is given by the formula

$$p = H \left(\frac{7.6}{T_a} - \frac{7.9}{T_c} \right),$$

in which p = draft pressure, in inches of water;

H = height of chimney, in feet;

T_a and T_c = absolute temperatures of the outside air and of the chimney gases, respectively.

EXAMPLE.—What draft pressure will be produced by a chimney 120 ft. high, the temperature of the chimney gases being 600° F., and of the external air 60° F.?

SOLUTION.—Substituting in the formula,

$$p = 120 \left(\frac{7.6}{460+60} - \frac{7.9}{460+600} \right) = .859 \text{ in.}$$

To find the height of chimney to give a specified draft pressure, the preceding formula may be transformed. Thus,

$$H = \frac{p}{\left(\dfrac{7.6}{T_a} - \dfrac{7.9}{T_c} \right)}$$

EXAMPLE.—Required, the height of the chimney to produce a draft of 1⅛ in. of water, the temperature of the gases and of the external air being, respectively, 550° and 62° F.

SOLUTION.—Substituting in the formula,

$$H = \frac{1.125}{\dfrac{7.6}{522} - \dfrac{7.9}{1,010}} = 167 \text{ ft}$$

In determiming the height of a chimney in cities, it should be borne in mind that it must almost always be carried to a height above the roofs of surrounding buildings, partly in order to prevent a nullification of the draft by opposing air-currents and partly to prevent the commission of a nuisance.

Area of Chimney.—The height of the chimney being decided on, its cross-sectional area must be sufficient to carry off readily the products of combustion. The following formulas for finding the dimensions of chimneys are in common use:

Let H = height of chimney, in feet;

$H. P.$ = horsepower of boiler or boilers;

A = actual area of chimney, in square feet;

E = effective area of chimney, in square feet;

S = side of square chimney, in inches;

d = diameter of round chimney, in inches.

Then

$$E = \frac{.3\,H.\,P.}{\sqrt{H}} = A - .6\sqrt{A} \qquad (1)$$

$$H. P. = 3.33\ E\ \sqrt{H} \qquad (2)$$

$$S = 12\sqrt{E} + 4 \qquad (3)$$

$$d = 13.54\ \sqrt{E} + 4 \qquad (4)$$

The accompanying table has been computed from these formulas.

EXAMPLE 1.—What should be the diameter of a chimney 100 ft. high that furnishes draft for a 600-H. P. boiler ?

SOLUTION.—Substituting in formula 1, $E = \dfrac{.3 \times 600}{\sqrt{100}} = 18.$

Now, using formula 4, $d = 13.54\sqrt{18} + 4 = 61.44$ in.

EXAMPLE 2.—For what horsepower of boilers will a chimney 64 in. sq. and 125 ft. high furnish draft ?

SOLUTION.—By simply referring to the table, the horsepower is found to be 934.

Maximum Combustion Rate.—The maximum rates of combustion attainable under natural draft are given by the following formulas, which have been deduced from the experiments of Isherwood:

Let F = weight, in pounds of coal per hour per square foot of grate area;

H = height, in feet, of chimney or stack.

Then, for anthracite burned under the most favorable conditions,

$$F = 2\sqrt{H} - 1 \qquad (1)$$

and under ordinary conditions,

$$F = 1.5\sqrt{H} - 1 \qquad (2)$$

For best semianthracite and bituminous coals,

$$F = 2.25\sqrt{H} \qquad (3)$$

and for less valuable soft coals,

$$F = 3\sqrt{H} \qquad (4)$$

SIZE OF CHIMNEYS AND HORSEPOWER OF BOILERS

Diameter (Inches)	Side of Square (Inches)	Actual Area (Square Feet)	Effective Area (Square Feet)	Commercial Horsepower — Height of Chimney, in Feet										
				50	60	70	80	90	100	110	125	150	175	200
18	16	1.77	.97	23	25	27								
21	19	2.41	1.47	35	38	41								
24	22	3.14	2.08	49	54	58	62							
27	24	3.98	2.78	65	72	78	83							
30	27	4.91	3.58	84	92	100	107	113						
33	30	5.94	4.47		115	125	133	141						
36	32	7.07	5.47			152	163	173	182					
39	35	8.30	6.57			183	196	208	219					
42	38	9.62	7.76			216	231	245	258	271				
48	43	12.57	10.44				311	330	348	365	389			
54	48	15.90	13.51				402	427	449	472	503	551		
60	54	19.64	16.98				505	539	565	593	632	692	748	
66	59	23.76	20.83					658	694	728	776	849	918	981
72	64	28.27	25.08					792	835	876	934	1,023	1,105	1,181
78	70	33.18	29.73						995	1,038	1,107	1,212	1,310	1,400
84	75	38.48	34.76						1,163	1,214	1,294	1,418	1,531	1,637
90	80	44.18	40.19						1,344	1,415	1,496	1,639	1,770	1,893
96	86	50.27	46.01						1,537	1,616	1,720	1,876	2,027	2,167

The maximum weight 'of combustion is thus fixed by the height of the chimney; the minimum rate may be anything less.

EXAMPLE.—Under ordinary conditions, what is the maximum rate of combustion of anthracite coal if the chimney is 120 ft. high?

SOLUTION.—By formula 2,

$$F = 1.5\sqrt{120} - 1 = 15\tfrac{1}{2} \text{ lb. per sq. ft. per hr.}$$

BOILER FITTINGS

Types of **Safety Valve.**—The *safety valve* is a device attached to the boiler to prevent the steam pressure from rising above a certain point. When steam is made more rapidly than it is used, its pressure must necessarily rise; and if no means of escape is provided for it, the result must be an explosion. Briefly described, the safety valve consists of a plate, or disk, fitting over a hole in the boiler shell and held to its place by a dead weight, by a weight on a lever, or by a spring. The weight or the spring is so adjusted that when the steam reaches the desired pressure the disk is raised from its seat, and the surplus steam escapes through the opening in the shell.

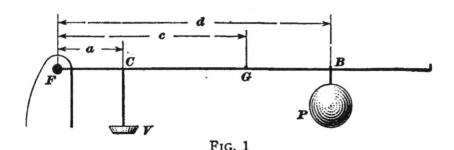

FIG. 1

Weight of Ball for Lever Safety Valve.—A simple diagram of a lever safety valve is shown in Fig. 1. The valve stem and the ball are attached to the lever at *C* and B, respectively, and the fulcrum is at *F*.

Let $d = F\,B =$ distance from fulcrum to weight, in inches;

　$c = F\,G =$ distance from fulcrum to center of gravity of lever, in inches;

　$a = F\,C =$ distance from fulcrum to center line of valve, in inches;

　$A =$ area of orifice beneath bottom of valve, in square inches;

　$W =$ weight of ball P, in pounds;

　$W_1 =$ weight of valve and stem, in pounds;

　$W_2 =$ weight of lever, in pounds;

　$p =$ blow-off pressure, in pounds per square inch.

Then, if the position of the ball P on the lever is fixed, the required weight of the ball may be found by the formula

$$W = \frac{a(pA - W_1) - W_2 c}{d}$$

EXAMPLE.—The area of the orifice is 10 sq. in., the distance from the valve to the fulcrum is 3 in., and the length of the lever is 32 in. The valve and stem weigh 5 lb., the lever weighs 12 lb., and the gauge pressure is 90 lb. What should be the weight W, if placed 2 in. from the end of the lever, assuming the lever to be straight?

SOLUTION.—In this case, $c = \frac{32}{2} = 16$ in., and $d = 32 - 2 = 30$ in. Then, substituting in the formula,

$$W = \frac{3(90 \times 10 - 5) - 12 \times 16}{30} = 83.1 \text{ lb.}$$

Position of Ball for Lever Safety Valve.—If the ball of a lever safety valve has a known weight and it is desired to find at what distance from the fulcrum it must be placed so as to give a required blow-off pressure, the formula to be used is

$$d = \frac{a(pA - W_1) - W_2 c}{W}$$

in which the various letters have the same meanings as before.

EXAMPLE.—Suppose all the quantities to remain the same as in the solution of the preceding example, except that it is desired that the boiler should blow off at 75 lb. gauge pressure, instead of 90 lb. What will be the distance of the weight from the fulcrum?

SOLUTION.—Applying the formula

$$d = \frac{\cdot 3(75 \times 10 - 5) - 12 \times 16}{83.1} = 24.58 \text{ in.}$$

Roper's Safety-Valve Rules.—Some inspectors of the United States Steamboat Inspection Service prefer to have lever safety-valve problems worked out by the rules that follow, known among American marine engineers as *Roper's rules.* A candidate for a marine engineer's license should always use Roper's rules when he knows that they are preferred by the examining inspector.

Let A = area of valve, in square inches;

   D = distance from center line of valve to fulcrum, in inches;

   L = distance of weight from fulcrum, in inches;

   P = steam pressure in pounds, per square inch;

   W = weight of load or weight on lever, in pounds;

   V = weight of valve and stem, in pounds;

   w = weight of lever, in pounds;

   l = distance from fulcrum to center of gravity of lever, in inches.

Then, the pressure at which the safety-valve will blow off is found by the formula

$$P = \frac{W\,L + w\,l + V\,D}{A\,D} \qquad (1)$$

If the distance L is known, the weight W to be hung on the lever is found by the formula

$$W = \frac{A\,P\,D - (w\,l + V\,D)}{L} \qquad (2)$$

The distance L from the fulcrum to the point at which the weight W is hung is found by the formula

$$L = \frac{A\,P\,D - (w\,l + V\,D)}{W} \qquad (3)$$

Area of Safety Valve.—By *area of safety valve* is meant the area of the opening in the valve seat; that is, the area of the surface of the valve in contact with steam when the valve is closed. The size of the valve relative to the size of the boiler and the working pressure is prescribed by law in many localities, and must be made to conform to the law wherever such law is in existence. In localities having no law governing this matter, the size of the safety valve may be calculated by the accompanying formulas, which are based on practice and recommended by leading authorities.

For natural draft,

$$A = \frac{22.5\ G}{p+8.62}, \qquad (1)$$

For artificial draft,

$$A = \frac{1.406\ w}{p+8.62}, \qquad (2)$$

in which G = grate surface, in square feet;

p = steam pressure, gauge, in pounds per square inch;

w = weight of coal burned per hour in pounds;

A = least area of safety valve, in square inches.

Location of Safety Valve.—The safety valve should be placed in direct connection with the boiler, so that there can be no possible chance of cutting off the communication between them. A stop-valve placed between the boiler and safety valve is a very fruitful cause of boiler explosions. Again, the safety valve must be free to act, and to prevent it from corroding fast to its seat, it should be lifted from the seat occasionally. Care must be taken to prevent persons ignorant of the importance of safety valves from raising the blow-off pressure by adding to the weights or increasing the tension of the spring. To this end, the weights of lever safety valves are often locked in position by the boiler inspector.

Use of Fusible Plugs.—*Fusible plugs* are devices placed in the crown sheets of furnaces or in similar places to obviate danger from overheating through lack of water. The plug often consists of an alloy of tin, lead, and bismuth, which melts at a comparatively low temperature. In many localities, the law requires that fusible plugs shall be attached to all high-pressure boilers.

Forms of Fusible Plugs.—The fusible plugs in common use are shown in section in Fig. 2. They consist of brass or iron shells threaded on the outside with a standard pipe thread. The plugs have some form of conical filling, the larger end of the filling receiving the steam pressure. The conical form of the filling prevents it from being blown out by the pressure of the steam. Fusible plugs applied from the outside differ from those applied from the inside, as Fig. 2 clearly shows.

Location of Fusible Plugs.—According to the rules issued by the Board of Boiler Rules of the State of Massachusetts, fusible plugs must be filled with pure tin, and the least diameter shall not be less than ⅓ in., except for working pressures over 175 lb., gauge, or when it is necessary to place a fusible plug in a tube, in which cases the least diameter of fusible metal shall not be less than ⅜ in. The location of fusible plugs shall be as follows:

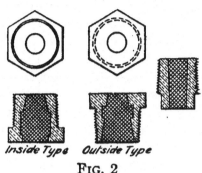

Inside Type *Outside Type*

FIG. 2

In horizontal return-tubular boilers, in the back head, not less than 2 in. above the upper row of tubes and projecting through the sheet not less than 1 in.

In horizontal flue boilers, in the back head, on a line with the highest part of the boiler exposed to the products of combustion, and projecting through the sheet not less than 1 in.

In locomotive-type or star water-tube boilers, in the highest part of the crown sheet and projecting through the sheet not less than 1 in.

In vertical fire-tube boilers, in an outside tube, placed not less than one-third the length of the tube above the lower tube-sheet.

In vertical submerged-tube boilers, in the upper tube-sheet.

In water-tube boilers of the Babcock & Wilcox type, in the upper drum, not less than 6 in. above the bottom of the drum and projecting through the sheet not less than 1 in.

In Stirling boilers of standard type, in the front side of the middle drum, not less than 6 in. above the bottom of the drum and projecting through the sheet not less than 1 in.

In Stirling boilers of the superheated type, in the front drum, not less than 6 in. above the bottom of the drum, and exposed to the products of combustion, projecting through the sheet not less than 1 in.

In water-tube boilers of the Heine type, in the front course of the drum, not less than 6 in. from the bottom of the drum, and projecting through the sheet not less than 1 in.

In **Robb-Mumford** boilers of standard type, in the bottom of the steam and water drum, 24 in. from the center of the rear neck, and projecting through the sheet not less than 1 in.

In water-tube boilers of the Almy type, in a tube directly exposed to the products of combustion.

In vertical boilers of the Climax or Hazelton type, in a tube or center drum, not less than one-half the height of the shell, measuring from the lowest circumferential seam.

In Cahall vertical water-tube boilers, in the inner sheet of the top drum, not less than 6 in. above the upper tube sheet.

In Scotch marine-type boilers, in the combustion-chamber top, and projecting through the sheet not less than 1 in.

In dry-back Scotch-type boilers, in the rear head, not less than 2 in. above the top row of tubes, and projecting through the sheet not less than 1 in.

In Economic-type boilers, in the rear head, above the upper row of tubes.

In cast-iron sectional heating boilers, in a section over and in direct contact with the products of combustion in the primary combustion chamber.

In other types and new designs, fusible plugs shall be placed at the lowest permissible water level, in the direct path of the products of combustion, as near the primary combustion chamber as possible.

Connection of Steam Gauge.—A steam gauge should be connected to the boiler in such a manner that it will neither be injured by heat nor indicate incorrectly the pressure to which it is subjected. To prevent injury from heat, a so-called siphon is placed between the gauge and the boiler. This siphon in a short time becomes filled with water of condensation, which protects the spring of the gauge from the injury the hot steam would cause. Care should be taken not to locate the steam-gauge pipe near the main steam outlet of the boiler, as this may cause the gauge to indicate a lower pressure than really exists. In locating the steam gauge, care must also be taken not to run the connecting pipe in such a manner that the accumulation of water in it will cause an extra pressure to be shown.

Bottom Blow-Off.—For the double purpose of emptying the boiler when necessary and of discharging the loose mud and sediment that collect from the feedwater, each boiler is provided with a pipe that enters the boiler at its lowest point. This pipe, which is provided with a valve or a cock, is commonly known as the *bottom blow-off*. The position of the blow-off pipe varies with the design of the boiler; in ordinary return-tubular boilers, it is usually led from the bottom of the rear end of the shell through the rear wall. Where the boiler is fitted with a mud-drum, the blow-off is attached to the drum.

Blow-Off Cocks and Valves.—While in many boiler plants globe valves are used on the blow-off pipe, their use is objectionable, because the valve may be kept from closing properly by a chip of incrustation or similar matter getting between the valve and its seat. As a result, the water may leak out of the boiler unnoticed.

Plug cocks packed with asbestos are widely used, the asbestos packing obviating the objectionable features of the ordinary plug cock. Gate valves are also used to some extent, but are open to the same objection as globe valves. In the best modern practice, the blow-off pipe is fitted with two shut-off devices. The one shut-off may be an asbestos-packed cock and the other some form of valve, or both may be cocks or valves, the idea underlying this practice being that leakage past the shut-off nearest the boiler will be arrested by the other.

Protection of Blow-Off Pipe.—When exposed to the gases of combustion, the bottom blow-off pipe should always be protected by a sleeve made of pipe, by being bricked in, or by a coil of plaited asbestos packing. If this precaution is neglected, the sediment and mud collecting in the pipe, in which there is no circulation, will rapidly become solid. The blow-off pipe should lead to some convenient place entirely removed from the boiler house and at a lower level than the boiler. Sometimes it may be connected to the nearest sewer. In many localities, however, ordinances prohibiting this practice are in force; the blow-off is then connected to a cooling tank, whence the water may be discharged into the sewer.

BOILER PIPING

Principal Considerations.—The piping of an engine and boiler plant requires that careful attention be paid to all the details as well as to the general design, not only in order to make it suitable for the purpose, but also in order to reduce the likelihood of a breakdown. The main considerations regarding steam piping are the size of the pipes; the arrangement and construction of the piping system; the method of providing for expansion; and proper drainage.

Materials for Pipes.—Most of the piping for steam and water is built up of wrought-iron or steel pipe of standard size. The various grades of wrought-iron and steel pipe are known as *standard, extra strong,* and *double extra strong.* Both wrought-iron and steel pipe are used in the piping systems of power plants. Formerly, wrought iron was chiefly used, but of late steel has been employed, especially for the larger sizes of steam pipes. The two kinds are equally reliable when made into expansion bends, copper bends as a general rule being used only for very heavy work.

Expansion Joints.—In installing steam piping, provision must be made for expansion and contraction, which ordinarily amounts to about 1½ in. per 100 ft. of pipe. Generally, this may be provided for in the arrangement of the piping; but for great lengths that are straight, or nearly so, it is necessary to use expansion joints, which may be made in various ways.

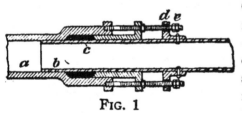

FIG. 1

One form, shown in Fig. 1, is called the *slip joint.* The ends *a* and *b* of the sections of pipe come together in a stuffingbox *c* in order to make a steam-tight joint. The stud bolts are extra long, so as to extend through holes in a flange *d* riveted to the pipe *b*. Check-nuts *e* on the ends of the studs prevent the two ends of the pipe from being forced apart by the steam pressure. The nuts *e* are not intended ordinarily to be in contact with the flange; their distance from the flange is adjusted so that the proper expansion may occur.

In Fig. 2 is shown a *corrugated expansion joint,* which is sometimes used on large exhaust pipes. It consists of a short
section of flanged corrugated
pipe, usually copper, which is
put in the steam pipe wherever
necessary. The elasticity of
this section, due to the corrugations, permits expansion and
contraction.

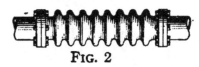

FIG. 2

Expansion Bends.—The best way of allowing for expansion is by using expansion bends, or bent pipes; but the space
they occupy often limits their use. The forms of bends more
commonly used are shown in Fig. 3, the trade name being given
below each bend. Where a bent pipe is used, the radius r of
the bend should not be less than six times the diameter of the
pipe, for wrought iron or steel; to secure the proper spring in
bends used on long lines of piping, the radius should be greater
than this. Bends of copper pipe may be of shorter radius,
as copper yields more readily than iron or steel.

Bends made from iron or steel pipe must be bent while red
hot. Iron and steel pipe bends generally have iron flanges
fastened on; copper bends either have composition flanges riveted and brazed on, or have steel flanges, the edges of the pipe
being turned over. The piping is usually installed so that it is
under a slight tension when cold; when filled with steam, the
expansion of the pipes removes the tension, and there is no
stress on the pipe except that due to the steam pressure.

Pipe Coverings.—To prevent loss of heat by radiation, steam
pipes are covered with various kinds of materials that are poor
conductors of heat. As a rule, the covering is manufactured
in short sections molded in halves to fit the pipe, the valves and
fittings being covered with the same material in a plastic state.
After the covering is properly secured in place, it is frequently
covered with a heavy duck or canvas jacket sewed on and
painted, and sometimes ornamented by brass bands placed at
regular intervals.

The substances used for covering steam pipes for this purpose are very numerous and vary considerably in efficiency.
Among the best and most widely used non-conducting materials
are hair felt, cork, magnesia, asbestos, and mineral wool.

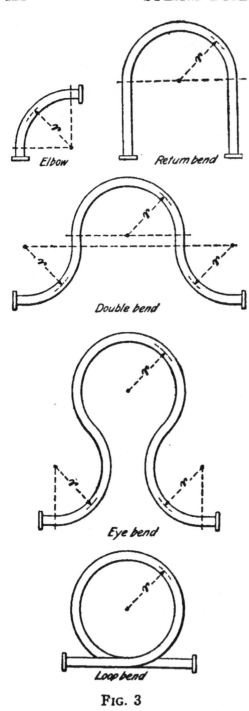

Elbow

Return bend

Double bend

Eye bend

Loop bend

FIG. 3

Frequently, pipe coverings are made up of combinations of two or more of these substances. For pipes laid in trenches, where a cheap covering is desired, such materials as sawdust, charcoal, coal ashes, coke, loam, and slaked lime are sometimes used.

Arrangement of Piping.—The pipes and fittings must be so proportioned as to permit a free flow of steam or water. Water pockets should be avoided; and where such pockets are unavoidable, they must be drained to free them from water. By-pass pipes should be arranged around feedwater heaters, economizers, pumps, etc. The system must be so designed as to give perfect freedom for expansion and contraction.

Perfect drainage must be provided in order that all water of condensation shall be fully separated from the steam. Reliability is insured by careful design and superior workmanship, combined with the use of high-class materials and fittings and the judicious placing of cut-out and by-pass valves. Drainage is best effected by arranging the piping so that all the water of condensation will flow by gravity toward a point close to the delivery end of the pipe, and then providing a drip pipe at that point. A trap may be placed at the end of the drip pipe for automatic draining.

BOILER FEEDING AND FEEDWATER

INJECTORS

Classification of Injectors.—Injectors may be divided into two general classes, namely, non-lifting and lifting injectors. *Non-lifting injectors* are intended for use where there is a head of water available. When the water comes to a non-lifting injector under pressure, as from a city main, it can be placed in almost any convenient position close to the boiler. *Lifting injectors* are of two distinct types, called *automatic injectors* and *positive injectors*. As positive injectors generally have two sets of tubes, they are frequently called *double-tube injectors*. Automatic injectors are so called from the fact that they will automatically start again in case the jet of water is broken by jarring or other means. Positive, or double-tube, injectors are provided with two sets of tubes, one set of which is used for lifting the water, and the other set for forcing the water thus delivered to it into the boiler. A positive injector has a wider range than an automatic injector and will handle a hotter feed-water supply; it will also lift water to a greater height than the automatic injector.

Size of Injector.—Most engineers prefer to select a size of injector having a capacity per hour about one-half greater than the maximum evaporation per hour in order to have some reserve capacity. The maximum evaporation, when not known, may be estimated in U. S. gallons by one of the following rules, which hold good for ordinary combustion rates under natural draft:

Rule I.—*For plain cylindrical boilers, multiply the product of the length and diameter in feet by 1.5.*

Rule II.—*For tubular boilers, either horizontal or vertical, multiply the product of the square of the diameter in feet and the length in feet by 1.9.*

Rule III.—*For water-tube boilers, multiply the heating surface in square feet by .4.*

Rule IV.—*For boilers not covered by the foregoing rules, multiply the grate surface in square feet by 12.*

Rule V.—*If the coal consumption in pounds per hour is known, it may be taken as representing the number of gallons evaporated per hour.*

No standard method of designating the size of an injector is followed by all makers; therefore, such an instrument must be selected from the lists of capacities published by the different makers.

Location of Injector.—An injector must always be placed in the position recommended by the maker. There must always be a stop-valve in the steam-supply pipe to the injector. While lifting injectors, when working as such, scarcely need a stop-valve in the suction pipe, it is advisable to supply it. When the water flows to the injector under pressure, a stop-valve in the water-supply pipe is a necessity. A stop-valve and a check-valve must be placed in the feed-delivery pipe, with the stop-valve next to the boiler. The check-valve should never be omitted, even if the injector itself is supplied with one. No valve should ever be placed in the overflow pipe, nor should the overflow be connected directly to the overflow pipe, but a funnel should be placed on the latter so that the water can be seen. This direction does not apply to the inspirator or to any other injector that has a hand-operated, separate overflow valve. In the inspirator, the overflow pipe is connected directly to the overflow, but the end of the pipe must be open to the air. In general, where the injector lifts water it is not advisable to have a foot-valve in the suction pipe, as it is desirable that the injector and pipe may drain themselves when not in use. A strainer should be placed on the end of the suction pipe.

Steam Supply to Injector.—The steam for the injector must be taken from the highest part of the boiler, as it must be supplied with dry steam. Under no consideration should the steam be taken from another steam pipe. The suction pipe should be as straight as possible and must be air-tight. In connecting up an injector, the pipes should be cleaned by blowing them out with steam before making the connection, because if a small bit of dirt gets into the injector it will interfere seriously with its operation.

Injector Troubles.—In the discussion of injector troubles as given in succeeding paragraphs, the suction pipe, strainer, feed-delivery pipe, and check-valve are considered as parts of the injector. In searching for the cause of a trouble, therefore, the suction and delivery pipes should be carefully inspected as well as the injector.

Failure to Raise Water.—The causes that prevent an injector from raising water are:

1. *Suction Pipe Stopped Up.*—This is due, generally, to a clogged strainer or to the pipe itself being stopped up at some point. In case the suction pipe is clogged, steam should be blown back through the pipe to force out the obstruction.

2. *Leaks in Suction Pipe.*—This prevents the injector forming the vacuum required to raise the water. To test the suction pipe for air leaks, plug up the end and turn the full steam pressure on the pipe; leaks will then be revealed by the steam issuing therefrom. Have the suction pipe full of water before steam is turned on, as the presence of small leaks will be revealed better by water than by steam.

3. *Water in the Suction Pipe Too Hot.*—A leaky steam valve or leaky boiler check-valve and leaky injector check-valve may be the cause of hot water or steam entering the source of supply and heating the water so hot that the injector refuses to handle it.

4. *Obstruction in Tubes.*—There may be an obstruction in the lifting or combining tubes; or, the spills (or openings) in the tubes through which the steam and water escape to the overflow may be clogged up with dirt or lime.

Injector Primes, but Will Not Force.—In some cases an injector will lift water, but will not force it into the boiler; or, it may force part of it into the boiler and the rest out of the overflow. When it fails to force, the trouble may be due to one or the other of the following causes:

10

1. *Choked Suction Pipe or Strainer.*—If the suction pipe or the strainer is partly choked, the injector, in either case, will be prevented from lifting sufficient water to condense all the steam issuing from the steam valve. The uncondensed steam, therefore, will gradually decrease the vacuum in the combining tube until it is reduced so much that the injector will not work. The remedy in case the supply valve is partly closed is to open it. In the case of a choked suction pipe, the obstruction should be blown out.

2. *Suction Pipe Leaking.*—The leak may not be sufficient entirely to prevent the injector from lifting water, but the quantity lifted may be insufficient to condense all the steam, which, therefore, destroys the vacuum in the combining tube. A slight leak may exist that will simply cut down the capacity of the injector. In such a case an automatic injector will work noisily, on account of the overflow valve seating and unseating itself as the pressure in the combining tube varies, due to the leak.

3. *Boiler Check-Valve Stuck Shut.*—If the boiler check-valve is completely closed, the injector may or may not continue to raise water and force it out of the overflow; this depends on the design of the injector. If the boiler check is partly open, the injector will force some of the water into the boiler and the remainder out of the overflow. In case the check-valve cannot be opened wide, water may be saved by throttling both steam and water until the overflow diminishes, or, if possible, ceases. The steam should be throttled at the valve in the boiler steam connection. If a check-valve sticks, it can sometimes be made to work again by tapping lightly on the cap or on the bottom of the valve body.

4. *Obstruction in Delivery Tube.*—Any obstruction in the delivery tube will cause a heavy waste of water from the overflow. To remedy this, the tube will have to be removed and cleaned.

5. *Leaky Overflow Valve.*—A leaky overflow valve is indicated by the boiler check chattering on its seat. To remedy this defect, grind the valve on its seat until it forms a tight joint.

6. *Injector Choked With Lime.*—It is essential to the proper working of an injector that the interior of the tubes be

perfectly smooth and of the proper bore. As in course of time they become clogged with lime, the capacity of the injector decreases until, finally, it refuses to work at all. If the water used is very bad, it becomes necessary frequently to cleanse the tubes of the accumulated lime. This may be done by putting the parts in a bath consisting of 1 part of muriatic acid to 10 parts of water. The tubes should be removed from it as soon as the gas bubbles cease to be given off.

Advantages and Disadvantages of Injectors.—The advantages of the injector as a boiler feeding apparatus are its cheapness, as compared with a pump of equal capacity; it occupies but little space; the repair bills are low, owing to the absence of moving parts; no exhaust piping is required, as with a steam pump; it delivers hot water to the boiler. The disadvantages of the injector are that it will not start with a steam pressure less than that for which it is designed, and that it will stand but little abuse, being poorly adapted for handling water containing grit or other matter liable to cut the nozzles.

INCRUSTATION AND CORROSION

Incrustation.—Broadly speaking, any deposit that is formed on the plates and tubes of a boiler is termed *scale*, or *incrustation;* it is caused by impurities that enter with the water and that are left behind in the boiler when the water is evaporated. In passing through the soil, water dissolves certain mineral substances, the most important of which are carbonate of lime and sulphate of lime. Other substances frequently present in small quantities are chloride of sodium, or common salt, and chloride of magnesium. It also often contains other troublesome substances.

Impurities in Feedwater.—Some of the more common impurities found in feedwater, together with their properties, are given in the following paragraphs:

Carbonate of lime will not dissolve in pure water, but will dissolve in water that contains carbonic-acid gas. It becomes insoluble and is precipitated in the solid form when the water is heated to about 212° F., the carbonic-acid gas being driven off by the heat.

Sulphate of lime dissolves readily in cold water, but not in hot water. It precipitates in the solid form when the water is heated to about 290° F., corresponding to a gauge pressure of 45 lb.

Chloride of sodium will not be precipitated by the action of heat unless the water has become saturated with it. Since it generally is present in but very small quantities in fresh water, it will take a very long time for the water in a boiler to become troublesome, and with the ordinary blowing down of a boiler once a week or every 2 wk., there is little danger of the water becoming saturated with it. Consequently, it is one of the least troublesome scale-forming substances contained in fresh water.

Chloride of magnesium is one of the worst impurities in water intended for boilers, for while not dangerous as long as the water is cold, it makes the water corrosive when heated, and when present in large quantities, it becomes dangerously corrosive, attacking the metal of the boiler and rapidly corroding it.

Organic matter by itself may or may not cause the water to become corrosive, but will often cause foaming; when it is present in small quantities in water containing carbonate or sulphate of lime, or both, it usually serves to keep the deposits from becoming hard.

Earthy matter, like organic matter, is not dissolved in the water, but is in mechanical suspension. It is very objectionable, especially when the earthy matter is clay, and when other scale-forming substances are present is liable to form a hard scale resembling Portland cement.

Acids, such as sulphuric acid, nitric acid, tannic acid, and acetic acid, are often present in the feedwater. The sulphuric acid is the most dangerous one of these acids, attacking the metal of which the boiler is composed and corroding it very rapidly. The other acids, while not so violent in their action as the sulphuric acid, are also dangerous, and water containing any one should be neutralized when it must be used.

Formation of Scale.—The small solid particles, due to precipitation of substances in solution or matter in mechanical suspension, remain for a time suspended in the water, especially the carbonate of lime that for some time after precipitation floats on the surface of the water. These particles will gradually settle on the plates, tubes, and other internal surfaces. A large part of the impurities will be carried by the circulation of the water to the most quiet part of the boiler and there settle and form a scale. In a few weeks, if no means of prevention are used, the inner parts of the boiler will be covered with a crust from $\frac{1}{16}$ to $\frac{1}{2}$ in. in thickness.

Danger of Scale.—A scale $\frac{1}{32}$ in. or less in thickness is thought by many to be an advantage, as it protects the plates from the corrosive action of acids in the water. When, however, the scale becomes $\frac{1}{4}$ in. thick or more, heat is transmitted through the plates and tubes with difficulty, more fuel is required, and there is danger of overheating the plates. The chief danger from a heavy incrustation is the danger of overheating the plates and tubes; it also prevents a proper examination of the inside of the boiler, since it may hide a dangerously corroded piece of plate or a defective rivet head that would otherwise be discovered.

Scale Containing Lime.—The carbonate of lime forms a soft, muddy scale, which when dry, becomes fluffy and flour-like. This scale may be easily swept or washed out of the boiler by a hose, provided it is not baked hard and fast. A carbonate scale is much harder to deal with when grease is allowed to enter the boiler. The grease settles and mixes with the floury scale, making a spongy crust that remains in contact with the plates, being too heavy to be carried off by the natural circulation of the water. The sulphate of lime forms a scale that soon bakes to the plates.

Kerosene as Scale Remover.—Some substances seem to soften and aid in detaching scale. Of these, kerosene oil has met with much favor. Its action appears to be mechanical rather than chemical, the oil penetrating or soaking through the scale and softening and loosening it. It is somewhat useful, too, in preventing the formation of scale, enveloping the fine particles of the scale-forming substances that, after precipitation, float on the surface of the water for a little while. It seems that this prevents the particles from adhering firmly to one another and to the metal when they finally settle.

Removal of Scale by Chipping.—A hard scale, when once formed, is generally removed by chipping it off with scaling hammers and scaling bars; soft scale can be largely removed during running by a periodic use of the bottom and surface blow-offs, and the remainder can usually be washed out and raked out when the boiler is blown down and opened. In order to prevent the scale-forming substances deposited on the metal from baking hard, it is advisable to let the boiler cool down slowly until entirely cold preparatory to blowing off, whenever circumstances permit this to be done. This cooling process will generally take from 24 to 36 hr.

Removal of Mud.—Mud and earthy matter by itself will not form any hard scale, but will often do so when carbonate of lime and sulphate of lime are present. An accumulation of such matter can be .prevented, and most of it can be removed, by a periodic use of the bottom blow-off, removing the remainder whenever the boiler is opened.

Internal Corrosion.—*Corrosion* of boiler plates may be defined as the eating away or wasting of the plates due to the chemical action of water. Corrosion may be internal and external.

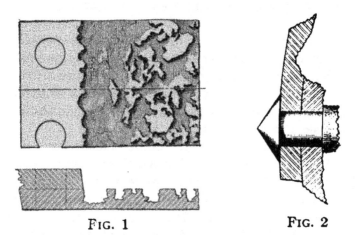

FIG. 1   FIG. 2

Internal corrosion may present itself as uniform corrosion, pitting or honeycombing, and grooving.

Uniform Corrosion.—In cases of uniform corrosion large areas of plate are attacked and eaten away. There is no sharp line of division between the corroded part and the sound plate.

Corrosion often violently attacks the staybolts and rivet heads.

Pitting or **Honeycombing.**—Pitting or honeycombing is readily perceived. The plates are indented in spots with holes and cavities from $\frac{1}{32}$ to $\frac{1}{4}$ in. deep. The appearance of a pitted plate is shown in Fig. 1. On the first appearance of pitting, the surface so affected should be thoroughly cleaned and a good coating of thick paint made of red lead and boiled linseed oil should be applied. This treatment should be given from time to time to insure protection to the metal.

Grooving.—Grooving, which means the formation of a distinct groove, is generally caused by the buckling action of the plates when under pressure. Thus, the ordinary longitudinal lap joint of a boiler slightly distorts the shell from a truly cylindrical form, and the steam pressure tends to bend the plates at the joint. This bending action is liable to start a small crack along the lap, which, being acted on by corrosive agents in the water, soon deepens into a groove, as shown in Fig. 2.

External Corrosion.—External corrosion frequently attacks stationary boilers, particularly those set in brickwork. The causes of external corrosion are dampness, exposure to weather, leakage from joints, moisture arising from the waste pipes or blow-off, etc. External corrosion should be prevented by keeping the boiler shell free from moisture, and the stoppage of all leaks as soon as they appear.

Leakage of rivets and the calking edges of seams may be caused by the delivery of the cold feedwater on to the hot plates; another cause is the practice of emptying the boiler when hot and then filling it with cold water. The leakage in both cases is due to the sudden contraction of the plates.

In horizontal water-tube boilers of the inclined-tube type, external corrosion principally attacks the ends of the tubes, especially the back ends, close up to the headers into which they are expanded. In the course of time the tubes will leak around the expanded portion in the headers.

If leaks are attended to as soon as they occur, no corrosion will take place, as the gases of combustion are harmless unless acting in conjunction with water or dampness, or unless the coal is rich in sulphur. Should, however, the ends of several tubes be found badly corroded but not yet leaking from that cause, the tubes should by all means be removed and replaced.

Lamination.—Sometimes what is called *lamination,* or the splitting of a plate into thin layers, is revealed by the action of the fire in causing a bag or blister to appear. Laminations due to slag and other impurities in the metal, which become

FIG. 3

flattened out when the plates are rolled, are shown at *a*, Fig. 3. Under the action of the heat the part exposed to the fire will form a blister, which may finally open at the point *b* or *c*. If the laminated portion of the plate is small, it may be cut out and a patch put in its place. If there are a number of laminations in the same plate, it is advisable to put in a new plate.

Overheating.—The heating of a plate beyond its normal temperature is called *overheating,* and may be caused by low water or by incrustation. When the plate is covered by a heavy scale, the plate becomes overheated, so that it yields to the steam pressure, forming a pocket, as shown in Fig. 4, which represents the shell sheet, or the sheet of a horizontal return-tubular boiler directly over the fire. If the pocket is

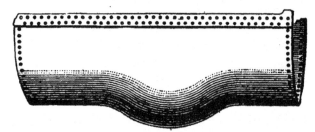

FIG. 4

not discovered in time for the plate to be repaired, it stretches until finally the material becomes too thin to withstand the steam pressure; the pocket then bursts with more or less liability of an explosion. The vegetable or animal oils carried into the boiler from a surface condenser are particularly liable to cause the formation of pockets.

Prevention of Incrustation and Corrosion.—Incrustation can best be prevented by purifying the feedwater prior to its entering the boiler, and can be fairly satisfactorily prevented by a chemical treatment of the water in case there is no purification of the water prior to its entering the boiler. When

SCALE-FORMING SUBSTANCES AND THEIR REMEDIES

Troublesome Substance	Trouble	Remedy or Palliation
Sediment, mud, clay, etc.	Incrustation	Filtration Blowing off
Readily soluble salts	Incrustation	Blowing off Heating feed
Bicarbonates of lime, magnesia, iron	Incrustation	Addition of caustic soda, lime, or magnesia
Sulphate of lime	Incrustation	Addition of carbonate of soda or barium chloride
Chloride and sulphate of magnesium	Corrosion	Addition of carbonate of soda, etc.
Carbonate of soda in large amounts	Priming	Addition of barium chloride
Acid (in mine water)	Corrosion	Alkali Heating feed
Dissolved carbonic acid and oxygen	Corrosion	Addition of caustic soda, slaked lime, etc.
Grease (from condensed water)	Corrosion	Slaked lime and filtering. Carbonate of soda Substitute mineral oil
Organic matter (sewage)	Priming	Precipitate with alum or chloride of iron and filter
Organic matter	Corrosion	Same as last

the water contains large quantities of substances that float on the surface, mechanical means may be resorted to, using the surface blow-off at frequent intervals or some equivalent skimming device. Corrosion is prevented by neutralizing the corrosive acids by an alkali; corrosion due to a perfectly fresh water can be prevented by giving a protective coating to the metal, which may be a thick red-lead paint made up with boiled linseed oil, or a thin coating of scale. Sometimes organic substances containing tannic acid, such as oak bark, hemlock, or sumac, are used to loosen or prevent scale. They are liable to injure the plates by corrosion and hence should not be used. The preceding table gives a list of troublesome scale-forming substances and the means of preventing or neutralizing them.

Use of Zinc in Boilers.—Zinc is much used in marine boilers for the prevention of both incrustation and corrosion. The scale may acquire thickness and hardness, but can easily be removed from the plates. The zinc is distributed through the boiler in the form of slabs. About 1 sq. in. of zinc surface should be supplied for every 50 lb. of water in the boiler.

TESTING OF FEEDWATER

Testing for Corrosiveness.—It is a good plan to test the feedwater and also the water in the boiler occasionally for corrosiveness. This may be done by placing a small quantity in a glass and adding a few drops of methyl orange. If the sample of water is acid, and hence corrosive, it will turn pink. If it is alkaline, and hence harmless, it will be yellow. The acidity may also be tested by dipping a strip of blue litmus paper in the water. If it turns red, the water is acid. This method is not so sensitive as the previous one, which should be used in preference. If litmus paper is kept in stock, it should be kept in a bottle with a glass stopper, as exposure to the atmosphere will cause the paper to deteriorate. If the water in the boilers has become corrosive and corrosion has set in, the water in the gauge glass will show red or even black. As soon as the color is beyond a dirty gray or straw color, it is advisable to introduce lime or soda to neutralize the acid.

Testing for Carbonate of Lime.—Pour some of the water to be tested into an ordinary tumbler. Add a little ammonia and ammonium oxalate, and then heat to the boiling point. If carbonate of lime is present, a precipitate will be formed.

Testing for Sulphate of Lime.—Pour some of the feedwater into a tumbler and add a few drops of hydrochloric acid.

Add a small quantity of a solution of barium chloride and slowly heat the mixture. If a white precipitate is formed, which will not redissolve when a little nitric acid is added, sulphate of lime is present.

Testing for Organic Matter.—Add a few drops of pure sulphuric acid to the sample of water. To this add enough of a pink-colored solution of potassium permanganate to make the whole mixture a faint rose color. If the solution retains its color after standing a few hours, no organic substances are present.

Testing for Matter in Mechanical Suspension.—Keep a tumblerful of the feedwater in a quiet place. If no sediment is formed in the bottom of the tumbler after standing for a day, there is no mechanically suspended matter in the water.

PURIFICATION OF FEEDWATER

Means of Purification.—Water intended for boilers may be purified by settlement, by filtration, by chemical means, and by heat. Filtration will remove impurities in mechanical suspension, such as oil and grease, and earthy matter, but will not remove substances dissolved in the water. Chemical treatment of the water will render the scale-forming substances and corrosive acids harmless, and may be applied either before or after the water enters the boilers, but preferablv the former. Purification by heat is based on the fact that most of the scale-forming substances become insoluble and precipitate when the water containing them in solution is heated to a high temperature.

Purification by Settlement.—For feedwater containing much matter in mechanical suspension, one of the simplest methods of purifying it is to provide a relatively large reservoir, or a large tank for small steam plants, where the impurities can settle to the bottom. While this method is fairly satisfactory, as far as earthy matter is concerned, it will not clear the water of finely divided organic matter, which is usually lighter than the water and often so finely divided as to be almost dissolved in it.

Purification by Filtration.—Organic and earthy matter in mechanical suspension is most satisfactorily removed by a filter, passing the water through layers of sand, gravel, hay, or equivalent substances, or through layers of cloth. Hay and cloth are of service especially where the feedwater contains oil or grease, as is the case where a surface condenser is used and the condensed steam is used over again.

Purification by Chemicals.—Chemical purification may take place before or after the water enters the boiler, the former method being somewhat more expensive. However, the purification is better carried out before the water enters the boiler for the reason that the amount of impurities entering the boiler will be greatly reduced. The chemical process to be adopted depends on the substances present in the water.

Use of Quicklime.—When the water contains only carbonate of lime, it may be treated with slaked quicklime, using 28 gr. of lime for every 50 gr. of carbonate of lime present in the water, the quicklime precipitating the carbonate of lime and being transformed into carbonate of lime itself during the process.

Use of Caustic Soda.—Water containing carbonate of lime may be treated with caustic soda, which precipitates the carbonate of lime and leaves carbonate of soda, which is harmless. For every 100 gr. of carbonate of lime 80 gr. of caustic soda should be added.

Use of Sal Ammoniac.—Sal ammoniac is sometimes added to water containing carbonate of lime and will cause the latter to precipitate. Its use is not advisable, however, on account of the danger of the formation of hydrochloric acid, which will attack the boiler. The formation of this acid is due to the use of an excessive quantity of sal ammoniac.

Treatment of Sulphate of Lime.—While slaked lime will precipitate carbonate of lime, it will have no effect on sulphate of lime, and water containing the latter, either alone or in conjunction with carbonate of lime, must be treated with other chemicals. The most available ones for water containing both are carbonate of soda and caustic soda. These are often fed into the boiler and will precipitate the carbonate of lime and sulphate of lime there, requiring the sediment to be blown out or otherwise removed periodically.

Quantity of Chemicals to Use.—When treating water containing carbonate of lime and sulphate of lime, caustic soda may be used either by itself or in combination with carbonate of soda, depending on the relative proportions of carbonate of lime and sulphate of lime present in the water. The amount of caustic soda or carbonate of soda to be used per gallon of feedwater can be found as follows:

Rule I.—*Multiply the number of grains of carbonate of lime per gallon by 1.36. If this product is greater than the number of grains of sulphate of lime per gallon, only caustic soda is to be used. To find the quantity of caustic soda required per gallon, multiply the number of grains of carbonate of lime in a gallon by .8.*

Rule II.—*Multiply the number of grains of carbonate of lime per gallon by 1.36. If this product is less than the number of grains of sulphate of lime per gallon, take the difference and multiply it by .78 to obtain the number of grains of carbonate of soda required per gallon. To find the amount of caustic soda required per gallon, multiply the number of grains of carbonate of lime in a gallon by .8.*

EXAMPLE.—A quantitative analysis of a certain feedwater shows it to contain 23 gr. of sulphate of lime and 14 gr. of carbonate of lime per gallon. How much caustic soda and carbonate of soda should be used per gallon to precipitate the scale-forming substances?

SOLUTION.—By rule I, $14 \times 1.36 = 19$ gr. As this product is less than the number of grains of sulphate of lime per gallon, rule II is to be used. Applying rule II, $(23 - 19) \times .78 = 3.12$ gr. of carbonate of soda, and $14 \times .8 = 11.2$ gr. of caustic soda.

Use of Carbonate of Soda.—Water containing sulphate of lime, but no carbonate of lime, may be treated with carbonate of soda. The amount of the latter that is required per gallon to precipitate the sulphate of lime is found by multiplying the number of grains per gallon by .78. When using soda, it is well to keep in mind that it will not remove deposited lime from the inside of a boiler. All that the soda can do is to facilitate the separating of the lime, that is, cause it to deposit in a soft state. This sediment must be removed periodically.

Use of **Trisodium Phosphate.**—For decomposing sulphate of lime, tribasic sodium phosphate, more commonly known as trisodium phosphate, is often used. This is claimed to act on the sulphate of lime, forming sulphate of sodium and phosphate of lime, the former of which remains soluble and is harmless, and the latter of which is a loose, easily removed deposit. Trisodium phosphate also acts on carbonate of lime and carbonate of magnesia, forming phosphate of lime and phosphate of magnesia, at the same time neutralizing the carbonic acid released from the carbonate of lime and magnesia, and the sulphuric acid released from the sulphates.

Neutralization of Acids.—Acid water can be neutralized by means of an alkali, soda probably being the best one. The amount of soda to be used can best be found by trial, adding soda until the water will turn red litmus paper blue.

Purification by Heat.—Carbonate of lime and sulphate of lime become insoluble if the water is heated, the former precipitating at about 212° F. and the latter at about 290° F. This fact is taken advantage of in devices that may be called combined feedwater heaters and purifiers; as they generally use live steam, they are also called *live-steam 'feedwater heaters.* Since no feedwater heater can effect a direct saving on fuel except when the heat is taken from a source of waste, it follows that a live-steam feedwater heater can affect the fuel consumption but indirectly. This it does by largely preventing the accumulation of scale in the boiler and the attendant loss in economy due to the lowering of the rate of heat transmission through a plate heavily covered with incrustation.

Economy of Heating Feedwater.—The feedwater furnished to steam boilers must of necessity be raised from its normal temperature to that of steam before evaporation can commence, and if not otherwise accomplished, it will be done at the expense of fuel that should be utilized in making steam. At 75 lb. gauge pressure the temperature of boiling water is about 320° F., and if 60° is taken as the average temperature of feedwater, $320 - 60 = 260$ B. T. U. is required to raise 1 lb. of water from 60° to 320°. It requires 1,151.5 B. T. U. to convert 1 lb. of water at 60° into steam at 75 lb. gauge pressure, so that the 260 B. T. U. required for heating the water represents

$260 \div 1{,}151.5 = 22.6\%$ of the total. All heat taken from a source of waste, therefore, that can be imparted to the feedwater before it enters the boilers is just so much saved, not only in cost of fuel but in boiler capacity.

Types of Exhaust-Steam Feedwater Heaters.—The impurities contained in the water will largely determine the type of exhaust-steam heater to be used in any given plant. These heaters are divided into two general classes, namely, open heaters and closed heaters.

An *open heater* is one in which the water space is open to the atmosphere. In a direct-contact open heater, the exhaust steam comes in contact with the water, which, by means of some one of a number of suitable devices, is broken into spray or thin sheets so that it will readily absorb the heat of the steam. In a coil heater, the exhaust steam passes through coils of pipe submerged in a vessel containing the water to be heated, and open at the top.

A *closed heater* is a heater in which the feedwater is not exposed to the atmosphere, but is subjected to the full boiler pressure. The steam does not come in contact with the water; the latter is heated through contact with metallic surfaces, generally those of tubes, that are heated by the exhaust steam.

Selection of Heater.—When the boiler feedwater is free from acids, salts, sulphates, and carbonates, so that no scale is formed at a high temperature, the closed feedwater heater will be found satisfactory. Heaters of the coil type may be used with pure water, but should not be used with water that will precipitate sediment or scale-forming matter of any kind. The coil heater is very efficient as a heater, as the water circulating through the coils is a long time in contact with the surface surrounded and heated by the exhaust steam. Heaters of the closed type, with straight tubes and sediment chamber, can be cleaned more readily than those having curved tubes, but the curved tubes allow more freedom for expansion and contraction. Heaters of the tubular type should have ample sediment chambers and may be used with water that contains organic or earthy matter, but not with water containing scale-forming ingredients. Carbonate of lime is likely to combine with earthy matter and form an exceedingly hard scale.

Heaters of the open exhaust-steam type have the advantage of bringing the exhaust steam in direct contact with the feed-water; some of the exhaust steam is condensed, thus effecting a saving in feedwater, and sediment and scale-forming ingredients, except sulphates of lime and magnesia, are precipitated or will settle to the bottom of the heater. The oil in the exhaust steam must be intercepted by special oil extractors, filters, or skimmers, generally combined with the heater and, by automatic regulation, sufficient fresh feedwater must be added to make up the total quantity required. When the system is properly arranged, all live-steam drips and discharges from traps are led to the heater.

BOILER TRIALS

Purposes of Boiler Trials.—A boiler trial, or boiler test, as it is often called, may be made for one or more of several purposes, the method of conducting the trial depending largely on its purpose. The boiler trial may vary from the simplest one, in which the only observations are the fuel burned and the water fed to the boiler in a stated period of time, to the elaborate standard boiler trial, in which special apparatus and several skilled observers are essential. The object of a boiler trial may be to determine the efficiency of the boiler under given conditions; the comparative value of different boilers working under the same conditions; the comparative value of fuel; or the evaporative power, or horsepower, of the boiler.

Observations **During Trial.**—The essential operations of a boiler trial are the weighing of the feedwater and fuel, and the observations of the steam pressure, temperature of feed-water, and various other less important pressures and temperatures. In conducting a boiler trial, the various observations of temperatures, pressures, etc. should be made simultaneously at intervals of about 15 min.

Weighing the Coal.—The coal supplied to the furnace is weighed out in lots of 500 or 600 lb. It is a convenient plan to have a box with one side open placed on a platform scale. A weight is then placed on the scale beam sufficient to balance the box. The scale may then be set at 500 or 600 lb., the coal shoveled in until the beam rises, and then fed directly from the box to the furnace. After the test, the ashes and clinkers must be raked from the ash-pit and grate and weighed. This weight subtracted from the weight of the coal used gives the amount of combustible.

Measurement of Feedwater.—The amount of water evaporated in a test for comparative fuel values may be taken as equal to the amount of feedwater supplied without introducing any serious error. The most reliable method of measuring the feedwater delivered to the boilers is to weigh it.

Standard of Boiler Horsepower.—When making a horsepower or an efficiency test, a more elaborate method of procedure is required than for a comparative fuel-value test. The reason for this is that different boilers generate steam at different pressures, different feedwater temperatures, and different degrees of dryness; hence, to compare the performances of boilers so as to determine their comparative efficiencies, it is necessary to reduce the actual evaporation to an *equivalent evaporation* from and at 212° F. per pound of combustible.

A committee of the American Society of Mechanical Engineers has recommended as a commercial horsepower *an evaporation of 30 lb. of water per hour from a feedwater temperature of 100° F. into steam at 70 lb. gauge pressure*, which is equivalent to 34½ units of evaporation; that is, to 34½ lb. of water evaporated from a feedwater temperature of 212° F. into steam at the same temperature.

Since 965.8 B. T. U. is required to evaporate a pound of water from and at 212°, a boiler horsepower is equal to 965.8 ×34½ =33,320 B. T. U. per hr.

Equivalent Evaporation.—The equivalent evaporation is readily determined by means of the formula

$$W_1 = \frac{W(H-t+32)}{965.8},$$

in which W = actual evaporation, in pounds of water per hour;

$\quad\quad H$ = total heat of steam above 32° F. at observed pressure of evaporation;

$\quad\quad t$ = observed feedwater temperature;

$\quad\quad W_1$ = equivalent evaporation, in pounds of water per hour, from and at 212° F.

11

EXAMPLE.—A boiler generates 2,200 lb. of dry steam per hour at a pressure of 120 lb. gauge. The temperature of the feed-water being 70° F., (*a*) what is the equivalent evaporation? (*b*) what is the horsepower of the boiler?

SOLUTION.—(*a*) According to the Steam Table, the total heat *H* corresponding to a gauge pressure of 120 lb. is 1,188.6 B.T.U. Applying the formula,

$$W_1 = \frac{2,200 \times (1,188.6 - 70 + 32)}{965.8} = 2,621 \text{ lb.}$$

(*b*) The horsepower, which is obtained by dividing the total equivalent evaporation by 34.5, the equivalent of 1 H. P., is,

$$2,621 \div 34.5 = 76 \text{ H. P., nearly}$$

Factor of Evaporation.—The quantity $\dfrac{H - t + 32}{965.8}$ that changes the actual evaporation of 1 lb. of water to equivalent evaporation from and at 212° F. is called the *factor of evaporation*. To facilitate the calculating of equivalent evaporation, the accompanying table of factors of evaporation is inserted. The equivalent evaporation is found by multiplying the actual evaporation by the factor of evaporation taken from the table.

EXAMPLE 1.—A boiler is required to furnish 1,800 lb. of steam per hour at a gauge pressure of 80 lb.; if the temperature of the feedwater is 48° F., what will be the rated horsepower of the boiler?

SOLUTION.—From the table, the factor of evaporation for 80-lb. pressure and a feedwater temperature of 40° is 1.214, and for the same pressure and a feedwater temperature of 50° it is 1.203; the difference is 1.214 − 1.203 = .011. The difference of temperature is 50° − 40° = 10°, and the difference between the lower temperature and the required temperature is 48° − 40° = 8°. Then, 10° : 8° = .011 : *x*, or *x* = .009; 1.214 − .009 = 1.205. 1,800 × 1.205 = 2,169 lb., and 2,169 ÷ 34.5 = 63 H. P., nearly.

EXAMPLE 2.—What is the factor of evaporation when the feedwater temperature is 122° F. and the gauge pressure 72?

SOLUTION.—In the table, under the column headed 70 and opposite 120 in the left-hand column is found 1.128; in column headed 80 and opposite 120 is found 1.131; difference is .003.

Gauge Pressures

Factors of Evaporation

Temperature of Feedwater Degrees F.	30	40	50	60	70	80	90	100	120	140	160	180	200
32	1.206	1.211	1.214	1.217	1.219	1.222	1.224	1.227	1.231	1.234	1.237	1.239	1.241
40	1.198	1.203	1.206	1.209	1.211	1.214	1.216	1.219	1.223	1.226	1.229	1.231	1.233
50	1.187	1.192	1.195	1.198	1.200	1.203	1.205	1.208	1.212	1.215	1.218	1.220	1.222
60	1.167	1.182	1.185	1.188	1.190	1.193	1.195	1.198	1.202	1.205	1.208	1.210	1.212
70	1.167	1.172	1.175	1.178	1.180	1.183	1.185	1.188	1.192	1.195	1.198	1.200	1.202
80	1.156	1.161	1.164	1.167	1.169	1.172	1.174	1.177	1.181	1.184	1.187	1.189	1.191
90	1.146	1.151	1.154	1.157	1.159	1.162	1.164	1.167	1.171	1.174	1.177	1.179	1.181
100	1.136	1.141	1.144	1.147	1.149	1.152	1.154	1.157	1.161	1.164	1.167	1.169	1.171
110	1.125	1.130	1.133	1.136	1.138	1.141	1.143	1.146	1.150	1.153	1.156	1.158	1.160
120	1.115	1.120	1.123	1.126	1.128	1.131	1.133	1.136	1.140	1.143	1.146	1.148	1.150
130	1.104	1.109	1.112	1.115	1.117	1.120	1.122	1.125	1.129	1.132	1.135	1.137	1.139
140	1.094	1.099	1.102	1.105	1.107	1.110	1.112	1.115	1.119	1.122	1.125	1.127	1.129
150	1.084	1.089	1.092	1.095	1.097	1.100	1.102	1.105	1.109	1.112	1.115	1.117	1.119
160	1.073	1.078	1.081	1.084	1.086	1.089	1.091	1.094	1.098	1.101	1.104	1.106	1.108
170	1.063	1.068	1.071	1.074	1.076	1.079	1.081	1.084	1.088	1.091	1.094	1.096	1.098
180	1.052	1.057	1.060	1.063	1.065	1.068	1.070	1.073	1.077	1.080	1.083	1.085	1.087
190	1.042	1.047	1.050	1.053	1.055	1.058	1.060	1.063	1.067	1.070	1.073	1.075	1.07
200	1.032	1.037	1.040	1.043	1.045	1.048	1.050	1.053	1.057	1.060	1.063	1.065	1.067
210	1.022	1.027	1.030	1.033	1.035	1.038	1.040	1.043	1.047	1.050	1.053	1.055	1.057

In the same vertical columns and opposite 130 are found **1.117**
and 1.120; difference is .003; same as above.   Hence, for an
increase of 10 lb. in gauge reading, there is an increase of .003
in the factor of evaporation, or an increase of .0003 for 1 lb.
and of .0003$\times$2 = .0006 for 2 lb.   Therefore, for a feedwater
temperature of 120° and 72 lb. pressure, the factor of evaporation is 1.128 + .0006 = 1.1286.   The difference between the
numbers opposite 120 and 130 in the two columns headed 70
and 80, respectively, is 1.128 − 1.117 = .011, and 1.131 − 1.120
= .011, showing that, for an increase of temperature in the
feedwater of 10°, there is a decrease in the factor of .011, and
for 1° a decrease of .0011, or for 2° of .0022.   Hence, the value
of the factor for a temperature of 122° and a gauge pressure of
72 lb. is 1.1286 − .0022 = 1.126.

Boiler Efficiency.—The *efficiency* of a boiler may be defined
as the ratio of the heat utilized in evaporating water to the
total heat supplied by the fuel.   The efficiency thus calculated
is really the combined efficiency of the furnace and boiler, as
it is not easily possible to determine separately the efficiency
of each.

The amount of heat supplied is determined by first accurately
weighing the fuel used during the test and deducting all the
ash and unconsumed portions.   This weight, in pounds, is
multiplied by the total heat of combustion of 1 lb. of the fuel,
as determined by an analysis, the product being the total
number of heat units supplied during the test under the
assumption that combustion was perfect.   The heat usefully
expended in evaporating water is obtained by first weighing
the feedwater and correcting this weight according to the
quality of the steam; the corrected weight is then multiplied
by the number of heat units required to change water at the
temperature of the feed into steam at the observed pressure.
The efficiency of a boiler, expressed in per cent., may be found
by the formula

$$E = \frac{A}{B},$$

in which E = efficiency of boiler;

A = heat utilized in evaporating water;

B = total heat supplied by fuel.

EXAMPLE.—A boiler trial shows a useful expenditure of 186,429,030 B. T. U. and a total supply of 270,187,000 B. T. U. What is the efficiency of the boiler?

SOLUTION.—Applying the formula,

$$E = \frac{186,429,030}{270,187,000} = .69 = 69\%$$

Standard Code.—For elaborate boiler trials, the standard code recommended by the American Society of Mechanical Engineers should be used.

BOILER MANAGEMENT

FILLING BOILERS

Preparation for Filling Boiler.—Before starting the flow of water into the boiler, the manhole plates or handhole plates that were removed preparatory to cleaning and overhauling must be replaced, and the blow-off valve must be closed. The gaskets, and also the surfaces with which they come in contact, should be examined to see that they are in good condition. It is customary to place a mixture of cylinder oil and graphite on the outer surface of each gasket, so that it may be removed without tearing. It is important that the manhole plates and handhole plates be properly replaced and secured in order to prevent leakage.

Height of Water.—In some cases the water can flow in and fill the boiler to the required height by means of the pressure that exists in the main supply pipe. In other cases, it may be necessary to use a hose or to fill the boiler with a steam pump or a hand pump. The boiler should be filled until the water shows half way up in the gauge glass.

Escape of Air.—While filling a boiler it is necessary to make provision for the escape of the contained air, since otherwise the pressure caused by the compression of the air may prevent the boiler from being filled to the proper height. Most boilers have some valve that can be used for this purpose; a gauge-cock may be left open until water issues therefrom, when it may be closed. Sometimes the manhole plate, if the manhole is on top, is left off while filling a boiler.

MANAGEMENT OF FIRES IN STARTING

Precautions in Starting.—After the boiler has been filled and before starting the fire, the attendant should see that the water column and connections are perfectly clear and free, that is, that the valves in the connections and the gauge-glass valves are open so that the water level may show in the glass; he should also see that the gauge-cocks are in good working order and open the top cock or the safety valve; he should take care that the stress on the stop-valve spindle is relieved by just unscrewing the valve from the seat without actually opening it. He should make sure that the pump, or injector, or whatever device is used to feed the boiler, is in good working order, and ready to start when required.

Starting the Fires.—It is customary to cover the grates with a layer of coal first, and then to add the wood, among which may be thrown oily waste or other combustible material that may be at hand. To start the fire, light the waste or other easily ignited material and open the damper and ashpit doors to produce draft. Then close the furnace door. After the wood has started to burn well, spread it evenly over the grate and add a fine sprinkling of coal, until this in turn begins to glow, when more coal may be added and the fire occasionally leveled until the proper thickness of fire has been obtained. It sometimes happens that the chimney refuses to draw; the draft can be generally started, however, by building a small fire in the base of the chimney.

Value of Slow Fires.—When getting up steam, the fire should not be forced, but, instead, should be allowed to burn up gradually. By forcing the fire, the plates or tubes that are nearest the fire suffer extreme expansion, while those parts that are remote from the fire are still cold; under such conditions the seams and rivets, and also the tube ends, which are expanded into the tube plates, are liable to be severely strained, and, possibly, permanently injured. It is not desirable to raise steam in any boiler, except in steam fire-engines, in less than from 2 to 4 hr., according to the size, from the time the fire is first started. When steam begins to issue from the opened top gauge-cock or the raised safety valve, as the case may be, the cock or the valve may be closed and the pressure still allowed to rise slowly until the desired pressure has been reached.

Trying the Fittings.—After the pressure at which the boiler is to run has been reached, and before cutting it into service, all the valves and cocks should be tried. The safety valve should be raised and its action noted; the water column should be blown out and the gauge-cocks tested; the feeding apparatus should be tried; and it should be noted particularly whether the check-valves seat properly and the valve in the feedpipe is open. All the accessible parts should be examined for leaks.

CONNECTING BOILERS

Cutting Boiler Into Service.—Cutting a boiler into service is accomplished by opening the stop-valve, thus permitting the steam to flow to the engine or other destination. The stop-valve should be opened very slowly to prevent a too sudden change in the temperature and consequent expansion of the piping through which the steam flows, and also to prevent water hammer. The steam-pipe drain should be kept open until the pipe is thoroughly warmed up. In large plants with many boilers and long steam mains it takes several hours to warm these pipes thoroughly by a slow circulation of the steam, but not until then should the main stop-valve be fully opened.

Connecting Boilers to Main.—Before connecting the different boilers of a battery to the same steam main, the precaution of equalizing the pressures in the different boilers must be observed in order to prevent a sudden rush of steam from one boiler to another. All the pressures should be equal within a variation of about 2 lb. before an attempt is made to connect the boilers.

Changing Over.—In plants where there are duplicate sets of boilers, one set being in operation while the other is undergoing repairs, overhauling, and cleaning, the method of *changing over*, or connecting, is as follows: Start the fires and raise steam in the boilers that are to be cut into service. Allow the pressure to rise in all to within 5 lb. of that which is in the boilers in operation. All arrangements before changing over should be made with a view to getting all the heat that can be obtained from the fires in the boilers that are to be cut out. This can be accomplished by running until the fires have given up all of their available heat for making steam, as indicated by the gradual fall in pressure when the dampers are wide open, and then making the change. While the fires in one set of boilers are burning low and the pressure is falling, the pressure in the boilers to be cut in is gradually rising and meeting, so to speak, the falling pressure of the set in operation. When the difference of 5 lb. is reached, change over. A man should be stationed at each stop-valve, and while one is being opened the other should be closed; the engine will continue running uninterruptedly while the change is being made.

EQUALIZING THE FEED

When the boilers of a battery have been cut into service and hence are all connected together through the steam main, the regulation and equalization of the feedwater becomes an important factor. Each boiler has its own check-valve and feed stop-valve, and generally all the boilers are supplied from one pump, which is running constantly. The quantity of water admitted to each boiler is regulated by its feed stop-valve. When the water gets low in any boiler, the feed stop-valve should be opened wider, while at the same time the feed stop-valves on one or more of the other boilers in operation may be closed partly and thus divert the feedwater to the one most requiring it. Some boiler plants have check-valves with an adjustable lift; in that case the feed is equalized by adjusting the lifts of the check-valves, the stop-valves being left wide open while running. It will be understood from the foregoing that the object in view is the maintaining of an equal water level in all the boilers through the manipulation of the feed stop-valves or check-valves. A boiler that is not doing its legitimate share in generating steam may be known by the fact that the feed stop-valve or check-valve on that boiler will be nearly, if not entirely, closed most of the time.

FIRING WITH SOLID FUEL

General Remarks.—The safe and economical operation of steam boilers calls for careful and intelligent management. The fires should be kept in such condition as to maintain the desired pressure and to burn the fuel with economy. Different fuels require different handling and hence only general rules can be given; much will depend on the skill and judgment of the attendant, who must himself discover in each case by actual trial the best method to pursue. The fires must be cleaned at intervals; the time and method of cleaning depend on conditions such as the nature of the fuel, the rapidity with which it is being consumed, the style of grate in use, and the construction of the furnace. Here much is left to the choice and judgment of the attendant, who should readily discover what is best to be done in any particular case.

Cleaning of Fires.—There are two methods employed in cleaning the fires: first, that of cleaning the front half and then the rear half; second, that of cleaning one side of the fire and then the other side. In the first method, previous to cleaning, green fuel is thrown on and allowed to burn partly until it glows over the entire surface. The new and glowing fuel is then pushed to the back of the furnace with a hoe, leaving nothing on the front half of the grate but the ashes and clinkers, which are then pulled out, leaving the front end of the grate entirely bare. The new fire which had been pushed back is now drawn forwards and spread over the bare half of the grate. The ashes and clinkers that are on the rear half of the grate are then pulled over the top of the front half of the fire and out through the furnace door; this leaves the rear half of the grate bare, which must be covered by pushing back some of the new front fire. The clean fire having been spread evenly, some new fuel must be spread over the entire surface.

The second method referred to is substantially the same in principle as that just described, with the difference that the fire is pushed to one side instead of to one end of the furnace, as in the first method described. The condition of the fires themselves and the nature of the service of the plant will determine just how often and at what time the cleaning of fires should take place. In general, the fires in stationary boilers require cleaning at intervals of from 8 to 12 hr. Fires require cleaning more often when forced draft is used than when working with natural draft.

Rapidity in Cleaning Fires.—Rapidity in cleaning fires is of great importance, as during the operation a large volume of cold air enters the furnace and chills the metallic surfaces with which it comes in contact; consequently, the boiler is damaged, however slightly. It is the greatest advantage of shaking grates that they allow the fire to be cleaned without opening the furnace door; the inrush of cold air and consequent chilling of the plates, etc. is thus avoided.

Before starting to clean fires, the steam pressure and the water level should be run up as high as is safe and the feed should be shut off in order to reduce the loss in pressure while cleaning. The condition of the fire during cleaning and the opening of the furnace doors cause the pressure to drop quite rapidly, but the rapidity and the amount of drop will be reduced by taking the precautions mentioned and cleaning quickly.

Drop of Pressure During Cleaning.—The amount of drop in pressure while cleaning fires depends on several conditions. For example, with a boiler that has a small steam space and, in addition, is too small for the work required of it without forcing, it is to be expected that the drop in pressure will be much more than if the reverse conditions exist. Furthermore, it may be necessary to clean fires while steam is being drawn from the boiler, instead of being able to clean at a time when the engine is stopped. In that case a greater drop must be expected than when cleaning while no steam is being drawn from the boiler. It is advisable when possible to do the cleaning at a time when no steam is being drawn from the boiler or when the demand for steam is light.

FIRING WITH LIQUID FUEL

Number of Oil Burners Required.—The number of oil burners to be installed in a given boiler depends on the type of burner to be used and the width of the furnace. The object to be attained is a uniform distribution of heat throughout the furnace. A straight-shot burner will produce a long, compact flame, whereas a fan-tailed burner will produce a wide, short flame; hence, one burner of the latter class may be sufficient for a furnace that would require two burners of the former class. One fan-tailed burner will ordinarily be sufficient for a furnace

6 ft. or less in width; two such burners may be used in furnaces from 6 to 14 ft. wide; and for furnaces over 14 ft. in width three burners should be installed. If narrow-flame burners are used, their number will have to be greater.

Location of Burners.—The burners may be installed in the centers of the fire-doors, if desired, and this practice is frequently followed. In case only one burner is employed, it may be placed in a suitable opening in the boiler front, between the fire-doors. Another method that has been used is to insert the burners below the fire-doors. So far as combustion is concerned, the location of the burner is of little importance, provided there is a sufficient air supply admitted under proper conditions; but the heat developed may be utilized to better advantage by exercising care in placing the burners. The flames should not be directed against the side walls, but should be approximately parallel thereto, with the tip of the burner about 6 or 8 in. above the brick covering of the grate. It is the usual practice to direct the flame from the front to the back of the furnace; but in some installations of water-tube boilers of the Stirling and Babcock & Wilcox types, the burners have been placed at the rear of the furnace, so as to direct the flames toward the front. This was found to be advantageous, as the combustion was completed in the portion of the furnace having the greatest cross-section, where the rapid expansion of the gases due to the heat generated could be accommodated most readily.

Necessity for Straining of Oil Fuel.—The crude oils used for fuel come from wells drilled in the earth, and as a result they contain varying proportions of sand and dirt. The denser and more viscous the oil, the greater is the tendency for it to retain impurities in suspension. If the oil is stored for some time in tanks, and is left undisturbed, some of the heavier dirt will settle to the bottom and thus be separated from the oil; but if the oil is taken direct from the wells to market, little or none of the dirt will be removed. Even fuel oil, which undergoes a preliminary heat treatment to separate the more volatile hydrocarbons from it, may not be free from foreign matter. It may be assumed, therefore, that all oil fuel contains dirt and sand, and as a consequence there arises the necessity of straining the oil before admitting it to the oil-burning system.   The presence of sand and grit in the oil supply will cause wear of the pump and erosion of the burner nozzles and orifices, and may result in the clogging of small orifices.

Oil Pressure.—One of the requirements for the efficient operation of an oil-burning plant is uniformity of the oil supply; for, if the oil supply is intermittent or varies in amount, the combustion will be irregular and incomplete.   To obtain a uniform rate of feed it is customary to supply the oil at a constant pressure by means of an ordinary duplex feed-pump. The pressure at which the oil is delivered to the burners varies in different plants and under different conditions.   It may range from 1 or 2 lb. to 150 lb. per sq. in., although the pressure is usually between 10 and 40 lb. per sq. in.   The type of burner employed will have some influence on the pressure required; also, forcing of the boiler above its normal rating will necessitate an increase in the oil pressure, to produce an increased flow of oil.

Objection to Gravity Feeding of Oil.—When a standpipe is used as a pressure regulator, the oil flows to the burners by gravity.   Some insurance companies refuse to insure plants in which gravity feeding is practiced, on the ground that, if a valve is left open, the boiler room may be flooded with oil, thus greatly increasing the danger from fire.   To overcome the objection to having a considerable amount of oil held in reserve above the level of the burners, a new form of apparatus has been produced.   This consists of a pump having a relief valve between the suction and delivery sides.   The relief valve is set at the desired oil pressure, so that, if that pressure is exceeded, the valve will lift and the excess of oil will be returned to the suction side of the pump.   When a standpipe is used, it should be fitted with an automatic drain valve that will open and drain out all the oil when the steam pressure falls below the lowest pressure at which the oil can be atomized. The flooding of the burners, with the attendant danger of explosion, will thus be averted.

Heating of Oil Fuel.—The viscosity of certain crude oils at ordinary atmospheric temperatures is very great, and it

Increases as the temperature becomes lower.   Below 40° F. the oil is so sluggish that it can scarcely be forced to the burners by the oil pump.   Consequently, in all localities where low temperatures are likely to occcur, provision must be made for heating oil fuel, so that it may flow readily through the pump and the pipes.   The flow of oil from the storage tank to the pump may be facilitated by surrounding the end of the suction pipe with a coil through which steam is led.   The oil in the vicinity of the pipe is thus heated and made more fluid. Oftentimes, the oil is heated after leaving the pump, to aid in obtaining better operation of the burners.   In any case, however, the heating must not be carried to a temperature sufficient to cause decomposition.   The temperature at which the decomposition begins depends on the nature and source of the oil.   Ordinarily the oil is not heated beyond a temperature of about 140° F.

Construction of Oil Tanks.—The size of the boiler plant will determine the character of the oil-storage tanks.   For plants of small or medium size, the tanks used are generally cylindrical in shape, built of steel plates, and coated with a protective covering of tar.   For large plants, rectangular tanks made of reinforced concrete are frequently used.   The penetrative properties of petroleum necessitate tight joints, and consequently it is advisable to entrust the construction of a steel tank to a boilermaker.   The rivet holes should be drilled rather than punched, so that the rivets will properly fill them, and the seams should be calked.   There should be no openings in the bottom, ends, or sides of the tank; all inlets and outlets should be in the top.   The manhole opening should be of such size as to permit ready entrance to the tank when required, and reinforcing flanges of steel or wrought iron should be riveted to the tank at the various openings.   Concrete tanks are usually made with partitions, so that deposits of sediment or of viscous matter may be removed at intervals without interfering with the continuity of the fuel supply.

Fittings for Oil Tank.—Even at ordinary temperatures, oil undergoes a slow process of evaporation, during which gases are evolved; consequently, every oil-storage tank should be fitted with a *ventilating pipe* to permit the escape of the gases thus set free. This pipe will also serve to lead off the air in the upper part of the tank during the operation of filling, and will thus prevent undue pressure from accumulating. The cross-sectional area of the ventilating pipe should be equal to that of the filling pipe. Care should be taken in locating the ventilating pipe to see that no naked light can approach its upper end; moreover, the openings in this end should be protected by a return elbow having wire gauze firmly fastened over it. The gases rising from the oil, when mixed with air in the correct ratio, form an explosive mixture, and could be ignited by a naked flame. This flame, however, would not pass through the fine meshes of the gauze, and the latter therefore forms a safety device; also, the downward curving of the elbow prevents any sparks from dropping into the opening. In addition to the ventilating pipe there should be a *telltale*, which is a device for indicating the amount of oil in the tank at any specified time.

Installation of Oil Tanks.—To conform with the requirements of underwriters and city ordinances, any oil-storage tank located above the surface of the ground should be at least 200 ft. from inflammable property. Moreover, the top of the tank should be at a lower level than the lowest pipe in the oil-burning system, so that, in case a valve is inadvertently left open, the plant will not be flooded with oil. If the tank is placed underground, as is usually the case, its top should be at least 2 ft. below the surface of the ground, and it should be 30 ft. distant from the nearest building; also, the top of the tank should not be at a higher level than the lowest pipe in the oil system. These precautions in locating the tank or tanks are necessary because of the highly inflammable nature of crude oil and fuel oil.

Separation of Water From Oil.—The crude oils invariably contain water in greater or less proportions. When the oil is run into the storage tank, the water, being heavier, sinks to the bottom of the tank and gradually accumulates. If the suction pipe extends to the bottom of the tank, some of this water will be drawn to the pump and forced to the burners, extinguishing the fires. To prevent this trouble, provision should be made for the removal of water from the oil tanks.

There is no practicable device that can be used to separate the water from the oil. The best thing to do is to let the water settle by gravity to the bottom of the tank· and then to pump it out at intervals, as required. The oil being somewhat lighter than water, will float on top of the latter, and by watching the discharge of the pump it will be easy to discontinue the pumping as soon as the flow of water ceases and oil appears.

Starting an Oil Fire.—Assuming that the oil pump has been put in operation and that the desired oil pressure has been obtained, the first step in starting an oil fire in a cold boiler is to open the damper. The valve in the oil supply pipe leading to the burner should be closed tightly. The needle valve, or oil-regulating valve, on the burner should be opened one turn, and the by-pass valve should be fully open. The steam valve is now opened, admitting steam to the burner, and the valve is left open until the steam blowing through appears dry. This operation heats up the burner, cleans the oil passages, and removes all water from the steam passages and pipes. The by-pass valve is now closed and the steam is almost, but not wholly, shut off. A bunch of oily waste is next lighted and thrown into the furnace, and the fire-door is closed. The oil-regulating valve is closed slightly, and the valve in the oil-supply pipe is opened fully. Steam and oil then pass through the burner and the spray is ignited by the burning waste. The condition of the fire is regulated by adjusting the steam valve and the oil-regulating valve, in conjunction with the ash-pit doors and the damper. The fire should not be forced, but should be increased slowly, so as to permit the boiler to accommodate itself to the increasing temperature.

The procedure just outlined is based on the assumption that a supply of steam for atomizing is available from an active boiler in the battery, or else from a small auxiliary boiler intended solely for the purpose. In case there is but a single boiler in the plant, the method of starting a fire will be somewhat different from that just described. The ash-pit doors and the fire-doors should first be opened, and a wood fire should be built in the furnace. This fire should be kept going until the gauge on the boiler shows a pressure of about 20 lb. Then the steam may be admitted to the burner and the latter may be started in the manner already described, using the wood fire to light the oil spray.   The wood fire should be built on the bottom of the furnace; or, in case it is a coal-burning furnace converted for oil burning, the fire should be built on the brick paving covering the grates.   It is not necessary to remove the bricks, nor must the burner be taken out while the wood fire is burning,   Care should be taken, however, to keep the fire about a foot from the burner, to prevent overheating of the latter.   After the oil burner has become well started, the wood fire may be raked out.

Furnace Conditions **With Oil Burning.**—When the brick-work of the furnace has become heated and the burner is working normally, the furnace space should appear to be filled with flame.   In every properly arranged oil-burning boiler there should be peep-holes at different points, to enable the fireman to determine the conditions of combustion in the furnace.   It is also advantageous to have the top of the chimney visible to the fireman, as the conditions at that point serve as a guide in the regulation of the fire.   The flame in the furnace should be white or golden white in color, and should be steady.   A properly adjusted steam burner will give a dazzling flame, whereas an air burner will produce a duller, yellower flame.   If the burner passages are not kept clean, or if the burner is improperly adjusted, the flame will become irregular and smoke will be produced.   No smoke should appear at the top of the chimney; instead, there should be a light, grayish haze when the burner is properly adjusted.

Smoke will be produced if there is too great a supply of oil, or too little air, or insufficient steam for atomization.   The remedies to be applied to rectify these faults of operation are obvious.   If the burner hisses or spits, it is probable that the steam used for atomizing contains moisture, or that there is water in the oil supplied to the burner, or that a leak has developed in the suction pipe, so that air is being forced into the burner with the oil.   As a rule, oil burning in steam-boiler furnaces is accompanied by a roaring noise of greater or less intensity, and firemen experienced in the burning of oil fuel are able to detect changes in the furnace conditions merely by the altered roaring of the burners.   An excessive supply of cold air will increase the noise greatly and at the same time will cause loss of heat; hence, as a general rule, it is safe to assume that, the quieter the fire, the more nearly perfect is the combustion. This is further corroborated by the fact that preheating of the air, which is conducive to better combustion, reduces the roaring. Sputtering of the flame from the burner may be due to the heating of oil to such a point as to cause it to be vaporized.

Draft Required for Oil Burning.—The draft of the chimney performs the function of drawing air into the furnace and of pushing the gaseous products of combustion through the boiler and out into the air. When solid fuel is used, a large part of the draft pressure is required to force the air through the layer of fuel on the grates, the remainder serving to overcome the resistance to the flow of the hot gases through the furnace, tubes, flues, and chimney. When oil is used, there is no resistance due to a fuel bed on a grate, and the draft pressure simply overcomes the resistance due to the flow of the air and hot gases; consequently, the draft pressure required is much less than for solid fuel. It follows, then, that when a boiler is changed to use liquid fuel instead of solid fuel, the chimney that was satisfactory for the latter gives too strong a draft for the former, and care must be exercised to prevent the admission of an excess of air. The draft pressure necessary for the burning of oil fuel ranges from .1 to .5 in. of water, the former corresponding to economical firing and the latter to firing with a large excess of air. If the boiler is overloaded, the draft must be increased above that required for normal working.

Formation of Soot.—An insufficient supply of air or an excessive feeding of oil will result in the formation of soot, which will by deposited on the heating surfaces and will reduce the efficiency of heat transmission from the hot gases to the water in the boiler, and may result in the overheating of some parts. If such accumulations occur, they must be cleaned away, to maintain the evaporative efficiency of the boiler. With careful management of fires, boilers burning oil have been run at full capacity for weeks without the formation of a troublesome amount of soot. On the other hand, with coal as a fuel, it would have been necessary to clean the tubes daily.

12

Shutting Down an Oil Burner.—To shut down an oil-burner that is in active operation in a furnace, the oil valve in the supply pipe should be closed. The steam valve should next be nearly closed, so that only a small amount of steam passes. The oil-regulating valve should then be opened a full turn, and the by-pass valve should be fully opened, after which steam should be turned on again by manipulating the steam valve. Steam will thus be discharged through the oil passages of the burner, and all oil in them will be blown out, thus preventing baking or carbonizing of the oil and the clogging that would otherwise result. When the burner has been blown out, the by-pass valve should be closed, and finally the steam should be shut off completely. The burner will then be put out of action, but will be in condition to be started again at short notice.

Precautions in Relighting Fires.—If the fire should go out completely, the fireman should not open the fire-door to look for the cause of the trouble. His first act should be to shut off the oil, and this should be followed by shutting off of the steam. Then the furnace door may be opened, a piece of waste may be set on fire and thrown in, and the fire may be restarted in the usual way. The damper should be wide open when this is being done. When the fire goes out for only a moment, as may happen if a slug of water comes through the burner, it will generally be reignited by the heat of the incandescent walls. It is dangerous to leave the steam turned on and to increase the amount of oil fed, in order to restart the fire in an incandescent furnace; for the explosion at the instant of reignition may blow open the doors of the furnace, knock down the brickwork, or cause other damage. The safest method is to relight the fire with burning waste.

Accidental Oil Fires.—If oil should escape in quantities from the system and should become ignited, no attempt should be made to put out the fire by spraying it with water, as this will serve merely to spread the blazing oil and will make matters worse. Instead, sand or loose earth should be thrown on the burning oil, to smother the flames. In some plants, boxes of sand are kept at convenient points, in readiness for emergencies of this nature. Also, in some cases, steam pipes are run to the oil-storage tanks, so that, if the oil in the tanks should take fire, steam could quickly be run in to smother the blaze.

Thermal Advantages of Oil.—The calorific value of a pound of oil fuel is about 30% higher than that of a high-class coal, so that, by using oil instead of coal, the same amount of heat may be obtained with a smaller weight of fuel. As has already been shown, it is possible to obtain more nearly perfect combustion, using less excess of air, with oil than with coal, which increases the efficiency. Again, there is no repeated opening and closing of the fire doors when oil is used, and this prevents loss of heat and at the same time gives a better distribution of heat in the combustion chamber. Also, there is less soot deposited on the heating surfaces, in consequence of which the transfer of heat is rapid and there is less heat lost up the chimney. The capacity of the boiler may therefore be increased from one-third to one-half by changing from coal to oil, while for short periods the capacity may be doubled. Owing to the uniformity of combustion of oil, the metal of the boiler is not subjected to such severe conditions as when solid fuel is used.

Rapidity of Regulation.—Another great advantage of oil fuel is the quickness and ease with which the intensity of combustion may be altered. This is of particular importance in a plant that may be subjected to quick increases in the load. The fire in the furnace of an oil-burning boiler may be brought very quickly from a moderate heat to a most intense heat, to meet a sudden demand for more steam. Also, in case the load falls off very suddenly, the burners may be adjusted rapidly to produce a correspondingly smaller quantity of heat. In emergencies, the fires may be put out instantly, and may be almost as quickly relighted when the danger is past. The closeness with which the combustion may be made to follow the demand for steam enables an almost uniform steam pressure to be maintained by the use of automatic regulators for the oil pressure and air supply.

Economy of Storage and Handling.—Oil fuel may be stored and handled with less labor and at less cost than is possible with coal. The volume occupied by a given weight of oil is less than the volume of an equal weight of coal; also, because of the greater calorific value of oil, it is possible to store 50%

more heating value, in the form of oil fuel, in a given space, than can be done if coal is the fuel. Oil possesses the additional advantage that it does not deteriorate or lose its heating value when stored for some time. Coal, on the other hand, not only deteriorates but is liable to spontaneous combustion. The cost of handling oil is less than that of handling coal, for the reason that oil is run into the storage tanks by gravity, or else is pumped into and out of storage.

Saving of Labor and Equipment.—When oil is employed for fuel, instead of coal, there is no necessity for cleaning fires, and consequently the boiler can be operated continuously at maximum capacity. There is a lower temperature in the boiler room, because the fire-doors are kept shut. There is less wear and tear on the pumps and other machinery that may be installed in the boiler room, inasmuch as the absence of coal dust and ashes enables the boiler room to be kept clean. The expense of removing ashes is avoided, and there is a great saving in labor, as fewer firemen and attendants are necessary. There is no formation of clinker on the grates or side walls, and no firing tools are used, so that the damage to furnace linings by careless handling of tools is obviated.

Disadvantages of Oil Fuel.—One of the disadvantages of oil as a boiler fuel is its low flash point; however, if oil having a flash point of not less than 140° F. is used with care and judgment by firemen of ordinary intelligence, there should be no serious danger. The regulations as to the location of storage tanks for oil fuel may be found irksome; for, in the case of a plant situated in a thickly populated district in a city, it may be wholly impossible to place the storage tanks at least 30 ft. from the nearest building, and also underground. The temperature of the fire obtained from oil fuel is greater than that of a coal fire, and if the feedwater used contains much scale-forming matter, the intense heat may cause more rapid deposit of scale, and thus increase the cost of tube cleaning and repairs.

UNIFORM STEAM PRESSURE

Desirability of Uniform Pressure.—The attendant should aim to carry the pressure in the boiler as uniform as possible. The reason why a steady steam pressure and a steady water level are conducive to economy in the use of a fuel is to be found in the fact that with these conditions in a properly designed plant there will be a fairly steady temperature in the furnace, which, under normal conditions, is sufficiently high to insure a thorough ignition of the volatile matter in the coal. Now, with a constant demand for steam, a fluctuation in the steam pressure is caused by a change in the furnace temperature, assuming the feedwater supply to be constant, and whenever the steam pressure is down, the furnace temperature is low at the same time. In consequence of this, large quantities of the volatile matter in the coal often escape unconsumed and cause a serious loss of heat. Furthermore, with a steady steam pressure the stresses on the boiler are constant, and hence the life of the boiler will be increased and repair bills will be smaller than otherwise.

Maintenance of Uniform Pressure.—During the period of time between the cleaning of the fires, the pressure may be carried nearly uniform by observing the following instructions: Manipulate the feed apparatus so that just the necessary amount of water constantly enters the boiler and thus maintains a constant level. Intermittent feeding is practiced under certain local conditions, as, for example, where there is an injector or a pump that is so large that it would be impossible to run it continuously without increasing the height of the water level. In such a case, stop feeding just before firing; that is, do not feed while firing nor resume feeding until the new fire begins to make steam, as indicated by the rise of pressure on the gauge. If the pressure tends to rise above the standard or normal pressure, partly close the dampers and increase the quantity of feed, assuming in this case that no damper regulator is fitted and that hence the damper is regulated by hand. A damper regulator, systematic firing, and proper feeding are essential for carrying a practically uniform pressure. Should the pressure continue to rise, throw on more green fuel, close the damper, increase the feed, and only as a last resort open the furnace door.

A uniform steam pressure cannot be kept without proper firing. To maintain such a pressure the following directions should be observed: Keep the fire uniformly thick; allow no air holes in the bed of fuel. Fire evenly and regularly; be careful not to fire too much at a time. Keep the fire free from ashes and clinkers, and do not neglect the sides and corners while keeping the center clean. Do not, however, clean the fires oftener than is necessary. Keep the ash-pit clear.

Keeping Water Level Constant.—In connection with the maintenance of a constant water level, the following instructions should be followed: On starting to work, remember that the first duty of the fireman is to examine the water level. Try the gauge-cocks, as the gauge glass is not always reliable. If there is a battery of boilers, try the gauge-cocks on each boiler.

PRIMING AND FOAMING

Priming.—The phenomenon called *priming* is analogous to boiling over; the water is carried into the steam pipes and thence to the engine, where considerable damage is liable to take place if the priming is not checked in time. There are several causes for priming, of which the most common ones are the following: Insufficient boiler power; defective design of boiler; water level carried too high; irregular firing; and sudden opening of stop-valves.

When the boiler power is insufficient, the best remedy is to increase the boiler plant; the next best thing to do is to put in a separator, which, obviously, will only prevent the entrained water from reaching the engine, and will not stop the priming.

Defective design of a boiler generally consists of a steam space that is too small or a bad arrangement of the tubes, which may be spaced so close in an effort to obtain a large heating surface as to interfere seriously with the circulation. In horizontal return-tubular boilers, a sufficiently large steam space can be obtained by the addition of a steam drum; sometimes the top row of tubes can be taken out to advantage, which permits a lower water level. Defective circulation in horizontal fire-tube boilers is difficult to detect and to remedy; if it is due to a too close spacing of the tubes, a marked betterment may be effected by the removal of one or two vertical rows of tubes. The remedy for a water level that is too high is to carry the water at a lower level.

Evidences of **Priming.**—Priming manifests itself first by a peculiar clicking sound in the cylinder of the engine, due to water thrown against the heads. In cases of very violent priming, the water will suddenly rise several inches in the gauge glass, thus showing more water in the boiler than there really is. When priming takes place, it can be checked temporarily as follows: Close the damper, and thereby check the fires until the water is quiet; the engine stop-valve should also be partly closed to check the inrush of water. Observe whether the water drops in the gauge glass, and then, if more feed is needed, increase the feed. To prevent damage to the engine, open the cylinder drains. Regular and even firing tends to prevent priming.

Foaming.—The phenomenon called *foaming* is not the same as priming, though frequently considered so. Foaming is the result of dirty or greasy water in the boiler; the water foams and froths at the surface, but does not lift. A boiler may prime and foam simultaneously, but a foaming boiler does not always prime. Foaming while taking place is visible in the gauge glass and is best remedied by using the surface blow-off. If no surface blow-off is fitted, the bottom blow-off may be used in order to get rid of the dirty water. Like foaming, priming, will cause a wrong level to be shown, and hence the first thing to do in case of foaming is to quiet the water by checking the outrush of steam, either by slowing the engine down or by checking the fire, or by both.

SHUTTING DOWN AND STARTING UP

Preparations for Shutting Down.—Before shutting down for the night it is advisable to fill the boiler to the top of the glass, so as to be sure to have sufficient water to start with in the morning. The presence of possible leaks through the valves, tube ends, or seams necessitates this course of action. Even if no leaks exist, it is good practice to do this, if for no other reason than to admit of blowing out a portion before raising steam in the morning. All the gauge cocks should be tried and the water column should be blown out to insure their being free and clear.

Banking of Fires.—The fires may be banked at such a time that there will be about enough steam to finish the day's run, thus shutting down under a reduced pressure with only a remote possibility of its rising again through the night. If the fires are properly banked and the steam worked off while the feed is on, it will be remotely possible for the pressure to rise during the night to a dangerous extent. To bank the fires they should be shoved to the back of the grate and well covered with green fuel, leaving the front part of the grate bare, thus preventing any possibility that the banked fire will burn up through the night.

Closing Valves **and** Damper.—The steam stop-valve, feed stop-valve, whistle valve, and other steam valves should be closed; the valves at the top and bottom of the gauge glass also should be shut off to prevent loss of water, etc. in case the glass should break during the night. If there is a damper regulator, it should be so arranged that the damper may be left closed, but not quite tight, because a small opening must be left to permit the collecting gases from the banked fire to escape up the chimney; otherwise there is danger that the accumulated gas will ignite and cause an explosion. It is very important to take this precaution and also to make a mark by means of which the distance the damper is open can be ascertained at a glance. In fact, a damper should be so made that when shut to the full extent of its travel there will be still sufficient space around it to allow the gas to escape. The damper regulator should be rendered positively inoperative in any manner permitted by its design so that the damper when closed will remain in that position until connected properly by the attendant in the morning.

Starting the Fires.—On entering the boiler room in the morning, the quantity of water in the boiler should first be noted. The gauge glass and the gauge-cocks should be tried and the water level determined. After it has been found that the water is not too low, the banked fires may be pulled down and spread over the grates and allowed to burn up slowly, the damper regulator, if one is fitted, in the meantime having been connected.

Blowing Down.—While the fires are burning and before the pressure begins to rise, the blow-off cock or valve should be opened and the boiler blown down; that is, a small quantity of the water should be blown out. This should be done every morning, so that any impurities in mechanical suspension in the water that settled during the night may be blown out. Great care should be exercised while blowing down that too much water is not blown out; from 3 to 4 in. as shown by the gauge glass, is sufficient. Under no circumstances should the attendant leave the blow-off while it is open. Disaster to the boiler is liable to follow a disregard of this injunction. Next, all the valves, except the stop-valve, which were shut the night before should be opened and tried to see that they are free and in good working order.

BOILER INSPECTION

NATURE OF INSPECTION

The inspection of a boiler usually consists of an external examination of the complete structure, and of the setting if the boiler is externally fired, and an internal inspection. The examination of the boiler consists of an ocular inspection for visible defects, and a *hammer test* or sounding for hidden defects of plates, stays, braces, and other boiler parts. The hammer test is made by tapping the suspected parts with a light hammer and judging the existence and extent of defects from the sound produced by the hammer blow. If the examination discloses marked wear and tear, a series of calculations is often required to find the safe pressure that may be allowed on the worn parts, using such formulas or rules as laws, ordinances, and regulations may prescribe for the particular official inspectors. In the absence of officially prescribed formulas and rules, the inspector should use such rules as he deems best applicable or in best accordance with good practice. The inspection is usually, but not always, completed by a so-called *hydrostatic test*, which is generally prescribed by official regulations.

EXTERNAL INSPECTION

Preparation.—Before a boiler that has been in use can be inspected, it must be blown out and must be allowed to cool off.

As soon as the water has been removed, the manhole covers, handhole covers, and washout plugs should be taken out and all loose mud and scale washed out with a hose. If the boiler is externally fired, the tubes must be swept and the furnace, the ash-pit, the smokebox, and the space back of the bridge wall must be cleaned out. Any removable insulating covering that prevents the inspector from having free access to the exterior of the boiler, must be removed to the extent deemed necessary by him; it may even be necessary to take down some of the bricks of the setting.

Inspection of Externally Fired Boilers.—In the inspection of an externally fired fire-tube or flue boiler, the exterior is first examined. The seams are gone over inch by inch; the rivet heads and calking edges of the plates are carefully scrutinized for evidence of leaks; and possible cracks are looked for between the rivet heads, especially in the girth seams and on the under side of the boiler. The plates must also be examined for corrosion, bulges, blisters, and cracks. The heads are inspected for cracks between the tubes or flues, cracks in the flanges, leaky tubes, and leaks in the seams. The condition of the firebrick lining of the furnace and bridge and the top of the rear combustion chamber is noted while making the exterior examination of the under side of the boiler. Every defect that is found should be clearly marked. Attention must also be paid to the condition of the grate bars and their supports.

Inspection of Internally Fired Boilers.—The inspection of the shell and heads must be followed by examination of the firebox or furnace tubes or flues, and of the combustion chambers if these are fitted inside the boiler. In fireboxes, special attention must be paid to the crown sheet. The ends of the staybolts require close examination; if such ends are provided with nuts, these must be examined, as they are liable to loosen and are also liable to be burned off in time. Each stay bolt should be tested for breakage, which is done by holding a sledge against the outside end of the staybolt and striking the inner, or firebox, end with a light hammer; in making this test on the boilers of locomotives it is customary, when practical, to subject the boiler to an internal air pressure of from 40 to 50 lb. per sq. in. The internal pressure, by bulging the sheets, separates the ends of a broken staybolt, which renders it comparatively easy to find them by the hammer test.

Inspection of **New** Boilers.—As made in boiler shops, the external inspection of new boilers, whether they are internally or externally fired, and whether they are of the water-tube or the fire-tube type, usually consists of a thorough examination for visible defects and testing under water pressure to locate leaks, if any. If a new boiler subject to official inspection during construction successfully passes such a hydrostatic test as the regulations prescribe, it will usually be permitted the working pressure it was designed for, the design having been approved officially before construction. The working pressure will be reduced, however, if the inspection discloses poor workmanship.

In the external inspection of water-tube boilers that have been in use, the tubes that are exposed directly to the heat of the fire must be particularly well examined for evidence of overheating. The plugs or handholes placed in headers to permit the insertion of the tubes and the cleaning of them are inspected for leakage, and the headers are inspected for cracks. Steam drums and mud-drums should be examined as carefully and for the same defects as the shells of externally fired fire-tube boilers. The firebrick lining of the furnace, and the interior of the brick setting in general, as well as the baffle plates controlling the direction of flow of the gases of combustion, must be examined for cracks and any other defects. The external inspection of the setting can usually be made very rapidly, as everything is in plain sight.

INTERNAL INSPECTION

Preparation.—Before the internal inspection is begun all loose mud should be washed out with a hose. In a horizontal return-tubular boiler and flue boiler, the shell plates and heads should be examined for corrosion and pitting; if the boiler has longitudinal lap seams, these should be inspected at the inside calking edge for incipient grooving and cracks. All seams should be examined for cracks between the rivet holes. Obviously, if the boiler is scaled to an appreciable degree, the scale must be removed before inspection. The tubes or flues should be examined for pitting, as well as for uniform corrosion.

All braces should be inspected by sounding them with a hammer, and if they are attached by cotter pins, it should be seen to that these are firmly in place. All defects found should be marked; it is good practice to make a memorandum of them as well. If any of the bracing seems to have worn considerably, it should be measured at the smallest part in order that the safe working pressure thereon may be calculated afterwards. To determine to what thickness a plate attacked by uniform corrosion has been reduced the inspector will have one or more holes drilled through the plate in the worn part to enable him to measure the thickness. These holes are afterwards plugged, generally by tapping out and then screwing in a plug.

Inspection of Locomotive-Type Boilers.—In internally fired boilers of the firebox and locomotive type, particular attention must be paid to the crown bars, crown bolts, and sling stays, and in boilers having the crown sheet stayed by radial staybolts, to these. As a general rule the inspector can make only an ocular inspection of most of them, as they are beyond his reach; where the outer sheets of the firebox contain inspection or washout holes above the level of the crown sheet, a lighted candle tied or otherwise fastened to a stick can usually be introduced through these holes from the outside by a helper. In inspecting above the crown sheet, the inspector should look for mud between the crown sheet and crown bars and sight over the top of the bars to see if any have been bent. As the inspector can reach from the inside of the boiler only a few of the staybolts staying the sides of the firebox, he must rely on the hammer test applied from the inside of the furnace for finding broken staybolts.

Flues and Combustion Chambers.—In boilers having circular furnace flues and internal combustion chambers the top of the furnace flues must be carefully inspected for deposits of grease and scale, which are especially liable to be found if the feedwater is obtained from a surface condenser. Even a light deposit of grease on the furnace flue is liable to lead to overheating and subsequent collapse of the top. The tops of the combustion chambers, together with their supports, are usually easily inspected, there being ample space to reach every part.

Inspection of Vertical Boilers.—Vertical boilers as a general rule, except in the largest sizes, have no manhole to admit a person to the inside, and such internal inspection as is possible must be made through the handholes. Defects to be looked for are pitting and uniform corrosion of the shell and tubes near the usual water-line, and cracks in the heads between the tubes, the lower head being especially liable to show this injury.

HYDROSTATIC TEST

Value of Test.—While a hydrostatic test is usually demanded by boiler laws and official rules and regulations, it does not at all follow that a boiler that has successfully stood this test will be safe. The chief value of the test lies in showing leaks. Under no consideration should the test pressure be such as to strain the parts of the boiler beyond the elastic limit of the material.

Care in Making Test.—When applying the hydrostatic test, the escape of air from the boiler while filling it with water should be provided for, as by raising the safety valve. After the boiler is full, all outlets must be tightly closed and the safety valve so blocked that it cannot open; the pressure must then be pumped up very slowly and carefully, the gauge being watched for any drop of pressure, which denotes a sudden yielding of some part of the boiler. When the desired test pressure has been reached, the boiler is inspected for leaks, and if any are found they are marked. Before calking to stop a leak, the pressure must be left off by opening some convenient valve or cock. In cold weather, when subjecting the boiler to the hydrostatic test, it is customary to heat the water to a lukewarm temperature.

Use of Blank Flange.—If the boiler to be tested is one of a battery, and the others are to be in use while this one is being inspected and tested, it is unwise to rely on a closed main stop-valve to break communication with the other boilers. It is good practice to put a blank flange between the boiler to be tested and its main stop-valve.

INSPECTION OF FITTINGS

Inspection of Safety Valve.—The safety valve requires very careful inspection. If this valve is known to leak, it should be reseated and reground before the hydrostatic test is made. After a boiler passes the hydrostatic test, the clamp locking the safety valve is removed, and by running the pressure up once more, the point at which the safety valve opens can be noted by watching the steam gauge, which is supposed to have been tested and corrected. If the safety valve does not open at the working pressure allowed or opens too soon, it is readjusted. If the safety valve is locked by a seal, as is often required by official regulations, the seal is applied after adjustment of the valve.

Testing of Steam Gauge.—The steam gauge should be tested before the hydrostatic test, and at each inspection, with a so-called boiler inspector's testing outfit. If the gauge under test is more than 5% incorrect, most inspectors will condemn it, although some will condemn gauges showing a much smaller error. In most cases the gauge can be repaired at small expense by the makers.

Inspection of Water Gauge and Blow-Off.—The connections of water columns and water-gauge glasses require examination in order to see that they are clear throughout their whole length. The blow-off pipe also requires examination in order to see that it is clear.

BOILER EXPLOSIONS

As a boiler inspector is usually called on to investigate the circumstances of a boiler explosion, and to render an opinion concerning it, he must be familiar with the causes of such explosions. Boiler explosions are really due to overpressure of steam. This may occur because the boiler is not strong enough to carry safely the working pressure used, or because the pressure, through some cause, such as sticking or overloading of the safety valve, has been allowed to rise above the ultimate strength of the boiler. A boiler may be unfit to bear its working pressure, for any of the following reasons: defective design; defects in workmanship or material; corrosion, and wear and tear in general; and mismanagement in operation.

The common faults in design that have led to boiler explosions are: insufficient staying, the stays being too small or too few in number; the cutting away of the shell for the dome, manhole, and other mountings, without reinforcing the edge of the plate around the hole; fixing the boiler too rigidly in its setting, thus causing it to be fractured on account of unequal expansion; defective water circulation in a boiler, which may lead to excessive incrustation and thus indirectly to explosion; and a poorly designed feed apparatus or safety valve. Defects in workmanship and material may include the use of faulty material containing blisters, lamination, etc.; careless punching and shearing of plates; burning and breaking of rivets; burning or otherwise injuring the plates in flanging, bending or welding; scoring of the plates along the joints by sharp calking tools; and injury of the plates by the use of the drift pin. Old boilers may, while being patched with new plates, be injured by the operation of removing the old rivets and putting in new ones, and also by the greater expansion and contraction of the new plate as compared with the old plate. The strength of the shell may be weakened by corrosion, pitting, and grooving. In some exploded boilers, the plates have been found to have wasted to little more than the thickness of wrapping paper. Fractures that ultimately end in explosion may be produced by letting the cold feedwater come directly into contact with the hot plates.

If a boiler fractures while undergoing the hydrostatic test, the water escapes through the rent in the plate and no explosion takes place, because the cold water has little or no stored energy. But when a boiler filled with steam and water at a high temperature fractures, a violent explosion generally follows. The steam escaping through the opening diminishes the pressure, and, consequently, a new body of steam is formed from the water, which, by escaping, lowers the pressure still more, allowing the formation of another new body of steam at a lower pressure, and this operation is continued until the pressure reaches that of the atmosphere. The formation of several successive large bodies of steam in this way, which occurs almost instantaneously, produces a disastrous explosion. Generally speaking, the larger the body of the contained water, the more disastrous is the result. For this reason water-tube boilers, which contain a relatively small amount of water, are generally considered to be much safer than fire-tube boilers.

STEAM ENGINES

INDICATING OF ENGINES

Inside-Spring Indicator.—The *indicator* is an instrument that is used to determine the action of the steam in the cylin-

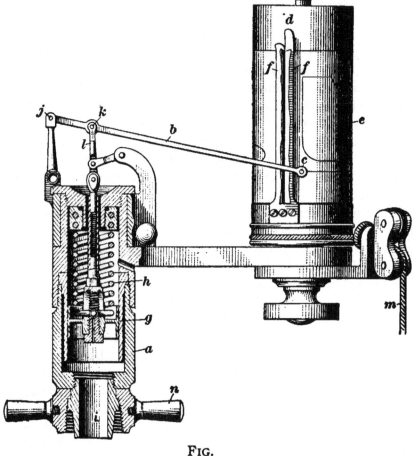

FIG.

der of an engine.   A form of indicator having its spring inside the barrel, or cylinder, is shown in Fig. 1.   The instrument consists essentially of a cylinder *a* containing a piston and a helical spring for measuring the steam pressure, a lever *b* for transmitting the motion of the piston to a pencil point *c*, and a drum *d* that carries paper on which this motion is recorded. The card *e* is held close to the drum by clips *f*, so that the pencil can easily trace the outline of the diagram. The piston, shown at *g*, must work in the cylinder as nearly frictionless as possible, the spring *h* being the only resistance to the upward motion of the piston. This spring is calibrated; that is, it is tested so as to determine the pressures required to move the pencil to various heights against the resistance of the spring. Hence, it is possible to find the pressure in the cylinder by the position of the pencil point. By turning a cock in the small pipe connecting the indicator with the engine cylinder, steam may be admitted to, or shut off from, the cylinder of the indicator at pleasure. When steam is admitted through the nipple *i*, its pressure causes the piston *g* to rise. The helical spring *h* is compressed, and resists the upward movement of the piston. The height to which the piston rises should then be in exact proportion to the pressure of the steam, and as the steam pressure rises and falls the piston must rise and fall accordingly.

To register this pressure, a pencil might simply be attached to the end of the piston rod, the point of the pencil being made to press against a piece of paper. It is desirable, however, to restrict the maximum travel of the piston to about ½ in. while the height of the card may advantageously be 2 in. To give a long range to the pencil while keeping the travel of the piston short, the pencil is attached at *c* to the long end of the lever *b*. The fulcrum of the lever is at *j*, and the piston rod is connected to it at *k* through the link *l*. The pencil motion is thus from four to six times the piston travel.

The indicator, however, not only must register pressures, but it must register them in relation to the position of the piston. This is accomplished by means of the cylindrical drum shown at *d*. This drum can be revolved on its axis by pulling the cord *m* that is coiled around it. When the pull is released, a spring on the drum spindle, inside the drum, turns the latter back to its original position. If the cord is connected with some part of the engine that has a motion proportional to the motion of the piston, the motion of the drum also will be proportional

13

to the motion of the piston. The outline then drawn by the pencil, termed an *indicator diagram*, shows the pressure on the piston at every point of the stroke. The slip of paper on which the diagram is drawn is called an *indicator card*.

Outside-Spring Indicator.—The spring of the indicator shown in Fig. 1 is subjected to the heat of the steam, and, when highly superheated steam is used in the engine, the spring may be rendered inaccurate by the heating due to the steam.

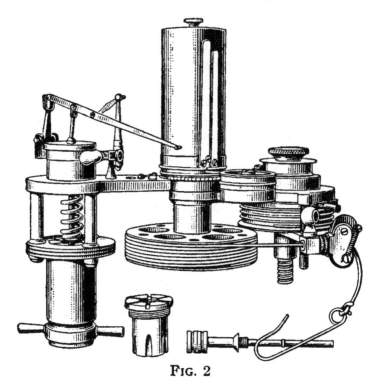

FIG. 2

An indicator with its spring located outside the barrel is shown in Fig. 2. This form obviates the danger of heating the spring and thus introducing errors, and at the same time it has the advantage of allowing the spring to be changed for a heavier or a lighter one, with less trouble. In other details this indicator is similar to that shown in Fig. 1.

Indicator Springs.—The height to which the piston will rise under a given steam pressure depends on the stiffness of the spring. Indicators are usually furnished with a number of springs of varying degrees of stiffness, which are distinguished by the numbers 20, 30, 40, etc. These numbers indicate the pressure, in pounds per square inch, required to raise the pencil 1 in. Thus, if a 40 spring is used, a pressure of 40 lb. per square inch raises the pencil 1 in., and therefore, the vertical scale of the diagram is 40 lb. per in. That is, the vertical distance, in inches, of any point on the diagram from the atmospheric line, multiplied by 40, gives the gauge pressure per square inch at that point. The scale of the spring chosen should not be less than half the boiler pressure, because it is not desirable to have the indicator card more than 2 in. in height.

Attachment of Indicator.—To attach the indicator to the engine, a hole is drilled in the clearance space of the cylinder

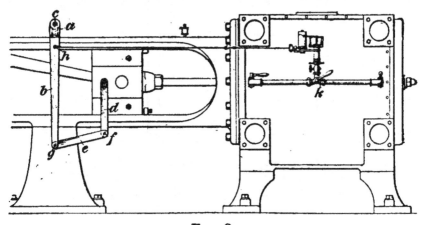

Fig. 3

and tapped for a ½-in. nipple, which should be as short as possible. The nipple has an elbow, into which is screwed a cock. The indicator may then be attached directly to the cock by the nut *n*, Fig. 1, the conical projection *i* of the indicator wedging tightly into the cock to prevent the leakage of steam. It is preferable to have an indicator at each end of the cylinder, but if that is not convenient, one indicator may be connected with both ends of the cylinder by means of a three-way cock, as shown at *k*, Fig. 3. This construction is undesirable, however, if the engine has a long stroke, as **the**

long pipes will cause considerable resistance to the flow of the
steam. The pipe connections between the indicator and the
engine should be short and direct, and care should be taken
to see that the piston at the end of the stroke does not cover
the hole tapped for the attachment of the indicator pipe.
Before attaching the indicator, it is advisable to open the cock
slightly, to blow out any dirt or rust that may have accumulated in the pipe.

Pendulum Reducing Motion.—The motion of the drum cord
is usually obtained from the crosshead. As the stroke of the

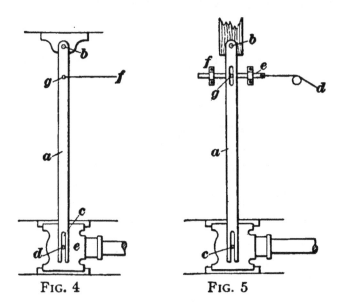

FIG. 4   FIG. 5

engine is nearly always greater than the circumference of the
drum, the cord cannot be attached directly to the crosshead,
and an arrangement called a *reducing motion* is used. A pendulum reducing motion is shown in Fig. 3. The upright *a* is
fastened to the engine frame, and the lever *b* is pivoted at *c*
to the upright. Another upright *d* is fastened to the crosshead or to the piston rod near the crosshead, and the link *e*
is connected at *f* to the piece *d* and at *g* to the lever *b*. The
cord, which should be parallel to the axis of the cylinder, is
attached to the point *h* on the lever *b*, which point must be
on the straight line connecting *c* and *g*.

Slotted-Lever **Reducing** Motions.—A form of reducing motion consisting of a swinging slotted lever is shown in Fig. 4. The lever *a* is pivoted at *b* to a stationary frame or girder and is slotted at *c* so as to fit over a pin *d* fastened to the crosshead *e.* The indicator cord *f* is attached to the lever at *g*.

Another form of slotted-lever reducing motion is shown in Fig. 5. The lever *a* is pivoted at *b* and is slotted at its lower end to fit over the pin *c* attached to the crosshead. The indicator cord *d* is fastened to a bar *e* that slides in the guides *f*.

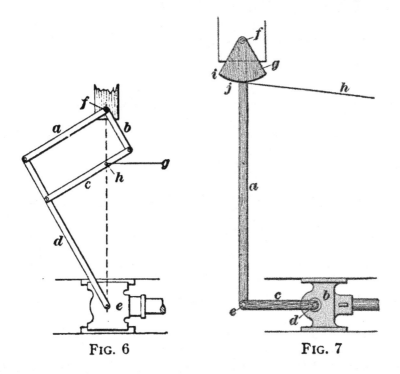

FIG. 6   FIG. 7

A pin *g* fixed in the bar *e* fits in a short slot in the upper end of the swinging lever. As the crosshead moves to and fro, the bar *e*, and hence the indicator cord also, copies the motion of the crosshead to a reduced scale.

Pantograph **Reducing** Motion.—The type of reducing motion shown in Fig. 6 is termed a *pantograph.* It consists of four straight bars *a*, *b*, *c*, *d* joined together with pin joints to form a parallelogram. One of the end bars *d* is prolonged, as shown, and is pivoted to the crosshead of the engine at *e*.

The uppermost corner of the parallelogram is pivoted at *f* to a stationary support, and the indicator cord *g* is attached to the bar *c* at the point *h*, which lies on the straight line joining the points *e* and *f*.

Brumbo Pulley.—A familiar form of reducing motion is that shown in Fig. 7, as it is easy to construct.   The lever *a* is connected with the crosshead *b* by the bar *c*, pivoted at *d* and *e*. At its upper end the lever *a* is pivoted on a stationary pin *f* and has firmly fixed to it the sector *g*.   The indicator cord *h* is fastened to the sector at the corner *i* and lies against the curved face *j* of the sector, which is an arc of a circle having

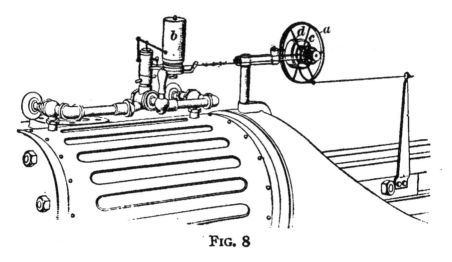

Fig. 8

its center at *f*.   As the crosshead moves back and forth, the cord *h* is given a similar but reduced. motion by the sector.

Reducing Wheels.—Instead of a reducing motion composed of levers, a *reducing wheel* may be used.   Such a device is illustrated in Fig. 8, as attached to the engine and to the indicator, ready for use.   A rigid upright is firmly fastened to the crosshead, and to this upright is tied a cord, the other end of which is wound on the wheel *a*.   As the crosshead moves back and forth, the cord rotates the wheel *a*.   Evidently the linear movement of a point on the rim of this wheel in any period is the same as that of the crosshead in that period. Fixed to the wheel *a* and turning with it on the same shaft is a smaller wheel *c,* on which is wound the cord leading to the indicator *b.* Hence, as the wheel *a* turns, the drum of the indicator is given a rotary motion that is proportional to the motion of the wheel *c,* and hence proportional to the crosshead movement also. But since the wheel *c* is so much smaller than wheel *a,* the movement of a point on the drum surface is much less than the movement of the crosshead. On the forward stroke, the wheel *a* is rotated against the resistance of a spring at *d;* but on the return stroke, this spring rotates the wheel in the opposite direction. Both wheels *a* and *c* are made as light as possible, in order that their inertia may not affect the accuracy of the reduction. The

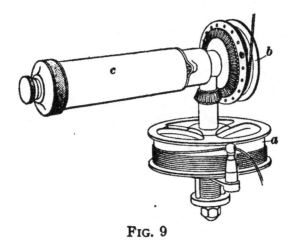

FIG. 9

cord leading from the wheel *a* to the upright on the crosshead must be parallel to the axis of the cylinder, but the cord from the wheel *c* to the indicator may incline upwards or downwards.

Reducing wheels, employing gears, are often made of aluminum for the sake of lightness. Such a wheel is shown in Fig. 9. It really consists of two wheels; on the larger one, shown at *a,* is wound the string that is attached to the arm on the crosshead, and from the smaller one *b* runs the cord to the indicator. A spring in the horizontal case *c* takes up the slack in the string. Frequently, the reducing wheel is attached directly to the body of the indicator, as shown in Fig. 2, thus avoiding the necessity of fastening it to the engine frame, as in Fig. 8.

Attachment of Indicator Cord.—The cord leading from a reducing motion to the indicator drum should be in two pieces with a hook on one of the free ends, preferably the end next to the indicator, and a loop in the end fastened to the reducing motion. This makes it possible to disconnect the indicator from the reducing motion when desired, and decreases the wear on the instrument. The length of the string should be carefully adjusted so as to give the drum the correct amount of motion. If the string is too short, it will be broken; and if too long, there will be lost motion and the card will not represent the true length of the engine stroke. It may also result in damage to the indicator.

A convenient arrangement is shown in Fig. 10. The hook *a* is attached to the indicator cord, and the cord *e* from the reducing motion is passed through a plate *b*, as shown. By

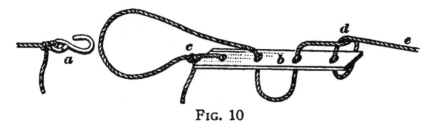

Fig. 10

slackening the cord at the point *d*, the plate may be slipped to any position along the cord. The length is thus easily adjusted. When the indicator is in operation, the hook is hooked into the loop. By unhooking the two cords the indicator may be stopped to put on a card.

The stretching of the indicator cord may introduce serious errors in the diagram. Hence it is better, if possible, to use a wire instead. If a cord is used, it should be as short as convenient. It should also be thoroughly stretched before being used.

Errors of Reducing Motions.—The forms of reducing motions shown in Figs. 3, 4, and 7 are imperfect, because the motion imparted to the cord is not exactly proportional to the movement of the crosshead. The only forms of reducing motion that are absolutely accurate are those in which the distance from the pivot to the point of attachment of the cord always bears a constant ratio to the distance from the pivot to the point where the lever is connected to the crosshead. This ratio must be the same at all points in the stroke, or at every position of the crosshead; if it is not, the reducing motion is not exact. In the reducing motion shown in Fig. 4, for instance, the distance from the pivot *b* to the center of the pin *d* is variable, depending on the position of the crosshead, whereas the distance from *b* to *g*, where the cord is attached, is always the same; in other words, the length of the long arm of the lever changes while the length of the short arm remains constant. As a result, the motions of the crosshead and the cord differ at different parts of the stroke, and the indicator diagram is correspondingly distorted. The reducing motions shown in Figs. 5 and 6, and the reducing wheels shown in Figs. 8 and 9, are accurate, inasmuch as the motion of the cord is at all times proportional to the motion of the crosshead.

Reduction of Errors.—For ordinary work with the indicator, the amount of error caused by the inexactness of reducing motions like those in Figs. 3, 4, and 7 is not serious and may be ignored. To secure a minimum of distortion of the diagram, the long lever should always be pivoted in such a position that it will be perpendicular to the line of movement of the crosshead when the latter is at the middle of its stroke. The accuracy of the motion will, in general, be increased by increasing the lengths of the long lever. For most purposes it is sufficient to use a lever whose length is twice the stroke of the engine.

Taking the Diagram.—After the instrument is properly attached, a blank card is slipped over the drum so as to fit smoothly, as in Fig. 1. The hook on the indicator cord is then engaged with the loop on the cord from the reducing motion, and the drum is allowed to rotate back and forth several times, to see that it works properly and that the cord is adjusted correctly. The cock is then opened and the indicator is allowed to work freely while the engine makes several revolutions. This warms up the parts to the working temperature. The pencil is then pressed lightly against the card during a single revolution. Next, the cock is closed and the pencil is again pressed against the card, recording the *atmospheric line.*

Finally, the cord is unhooked, and the card is removed from the drum.

If but one indicator and a three-way cock are used, as shown in Fig. 3, the cock is opened to admit steam from one end of the cylinder, and the diagram from that end is taken; then the cock is turned to admit steam from the other end, and another diagram is taken; finally, the steam is shut off entirely, and the atmospheric line is drawn.

CLEARANCE AND CUT OFF

Clearance.—The term clearance is used in two senses in connection with the steam engine. It may be the distance between the piston and the cylinder head when the piston is at the end of its stroke, or it may represent the volume between the piston and the valve when the engine is on dead center. To avoid confusion, the former is called *piston clearance*, and the latter is termed simply *clearance*. Piston clearance is always a measurement, expressed in parts of an inch. Clearance, however, is a volume.

The clearance of an engine may be found by putting the engine on a dead center and pouring in water until the space between the piston and the cylinder head, and the steam port leading into it, is filled. The volume of the water poured in is the clearance. The clearance may be expressed in cubic feet or cubic inches, but it is more convenient to express it as a percentage of the volume swept through by the piston. For example, suppose that the clearance volume of a $12''\times18''$ engine is found to be 128 cu. in. The volume swept through by the piston per stroke is $12^2\times.7854\times18=2{,}035.8$ cu. in. Then, the clearance is $128\div2{,}035.8=.063=6.3\%$. The clearance may be as low as $\frac{1}{2}\%$ in Corliss engines, and as high as 14% in high-speed engines.

Effects of Clearance.—Theoretically, there should be no clearance, since the steam that fills the clearance space does no work except during expansion; it is exhausted from the cylinder during the return stroke, and represents so much dead loss. This is remedied, to some extent, by compression. If the compression were carried up to the boiler pressure, there would be very little, if any, loss, since the steam would then fill the entire clearance space at boiler pressure, and the amount of fresh steam needed would be the volume displaced by the piston up to the point of cut-off, the same as if there were no clearance. In practice, however, the compression is made only sufficiently great to cushion the reciprocating parts and bring them to rest quietly.

It is not practicable to build an engine without any clearance, on account of the formation of water in the cylinder due to the condensation of steam, particularly when starting the engine. Automatic cut-off high-speed engines of the best design, with shaft governors, usually compress to about half the boiler pressure, and have a clearance of from 7 to 14%. Corliss engines require but very little compression, owing to their low rotative speeds; they also have very little clearance, since the ports are short and direct.

Apparent Cut-Off.—The *apparent cut-off* is the ratio between the portion of the stroke completed by the piston at the point of cut-off, and the total length of the stroke. For example, if the length of stroke is 48 in., and the steam is cut off from the cylinder just as the piston has completed 15 in. of the stroke, the apparent cut-off is $\frac{15}{48} = \frac{5}{16}$.

Real Cut-Off.—The *real cut-off* is the ratio between the volume of steam in the cylinder at the point of cut-off and the volume at the end of the stroke, both volumes including the clearance of the end of the cylinder in question. If the volume of steam in the cylinder, including the clearance, at the point of cut-off is 4 cu. ft., and the volume, including the clearance, at the end of the stroke is 6 cu. ft., the real cut-off is $\frac{4}{6} = \frac{2}{3}$.

Ratio of Expansion.—The *ratio of expansion*, also called the *real number of expansions*, is the ratio between the volume of steam, including the steam in the clearance space, at the end of the stroke, and the volume, including the clearance, at the point of cut-off. It is the reciprocal of the real cut-off. For example, if the volume at the end of the stroke is 8 cu. ft., and the cut-off is 5 cu. ft., the ratio of expansion is $8 \div 5 = 1.6$; in other words, the steam would be said to have one and six-tenths expansions. The corresponding real cut-off would be $\frac{5}{8}$.

Let e = real number of expansions;

i = clearance, expressed as a per cent. of the stroke;

k = real cut-off;

k_1 = apparent cut-off;

r = apparent number of expansions = $\dfrac{1}{k_1}$.

Then, $e = \dfrac{1}{k}$ and $k = \dfrac{1}{e}$ (1)

$$k = \frac{k_1 + i}{1 + i} \qquad (2)$$

EXAMPLE.—The length of stroke is 36 in.; the steam is cut off when the piston has completed 16 in. of the stroke; the clearance is 4%. Find the apparent cut-off, the real cut-off, and the real number of expansions.

SOLUTION.—Apparent cut-off = $\frac{16}{36} = \frac{4}{9} = .444$.

Real cut-off = $k = \dfrac{k_1 + i}{1 + i} = \dfrac{.444 + .04}{1 + .04} = \dfrac{.484}{1.04} = .465$.

Real number of expansions = $e = \dfrac{1}{k} = \dfrac{1}{.465} = 2.15$.

MEAN EFFECTIVE PRESSURE

Finding the Mean Effective Pressure.—In order to find the horsepower of an engine, it is necessary to know the *mean effective pressure,* abbreviated *M. E. P.,* which is defined as the average pressure urging the piston forwards during its entire stroke in one direction, less the pressure that resists its progress. The mean effective pressure is usually found from the indicator diagram in one of two ways.

1. The area of the diagram in square inches may be measured by an instrument called a *planimeter;* the *M. E. P.* is then found by dividing the area of the diagram in square inches by the length of the diagram in inches, and multiplying the quotient by the scale of the spring.

2. Where a planimeter is not available, the *M. E. P.* may be found with a fair degree of accuracy by multiplying the length of the mean ordinate by the scale of the spring.

Planimeter.—A common form of planimeter is shown in Fig. 1. It consists of two arms hinged together by a pivot joint at *j*. One arm carries a recording wheel *d*, which rolls on the surface to which the card is fastened, while the outline of the diagram is being traced by the point *f*. The needle point *p* is fixed in the paper or drawing board, and remains stationary during the operation.

The indicator card should be fastened to a smooth table or a drawing board that has previously been covered with a piece of heavy unglazed paper or cardboard. The point *p* should be placed far enough from the card to enable the wheel to roll on the unglazed paper without touching the card, as it will slip if rolled over a smooth surface. Set the zero of

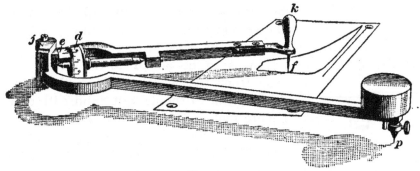

FIG. 1

the wheel *d* opposite the vernier *e;* then, with the tracing point *f*, follow the line of the diagram carefully, going around the diagram in the direction of the hands of a watch, and stop exactly at the starting point.

Reading the Vernier.—The area is read from the recording wheel and vernier as follows: The circumference of the wheel is divided into 10 equal spaces by long lines that are consecutively numbered from 0 to 9. Each of these spaces represents an area of 1 sq. in. and is subdivided into 10 equal spaces, each of which represents an area of .1 sq. in. Starting with the zero line of the wheel opposite the zero line of the vernier and moving the tracing point once around the diagram, the zero of the vernier will be opposite some point on the wheel; if it happens to be directly opposite one of the division lines on the wheel, that line gives the exact area in tenths of a square inch. The zero of the vernier, however, will probably be between two of the division lines on the wheel, in which case write down the inches and tenths that are to the left of the vernier zero, and from the vernier find the nearest hundredth of a square inch as follows: Find the line of the vernier that is exactly opposite one of the lines on the wheel. The number of spaces on the vernier between the vernier zero and this line is the number of hundredths of a square inch to be added to the inches and tenths read from the wheel. An example is presented in Fig. 2, where the 0 of the vernier lies between the lines on the wheel representing 4.7 and 4.8 sq. in., respectively, showing that the area is something more than 4.7 sq. in. Looking along the vernier it is seen that there are three spaces

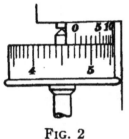

FIG. 2

between the vernier zero and the line of the vernier that coincides with one of the lines on the wheel; this shows that .03 sq. in. is to be added to the 4.7 sq. in. read from the wheel, making the area 4.73 sq. in., to the nearest hundredth of a square inch. The reading thus taken is the area of the diagram, in square inches. The M. E. P. is found by dividing this area by the length of the diagram on a line parallel with the atmospheric line, and multiplying by the scale of the spring.

EXAMPLE.—The area of the diagram is 4.73 sq. in., the length is 3.5 in., and a 40 spring is used; find the M. E. P.

SOLUTION.—M. E. P. $= \dfrac{4.73}{3.5} \times 40 = 54$ 1 lb. per sq. in.

Hints for Use of Planimeter.—It is well to place the fixed point p, Fig. 1, so that, as the tracing point moves around the diagram, the arms will swing about equally on each side of a position at right angles with each other. A slight dot is generally made with the tracing point to mark the point at which its motion around the diagram begins; when the tracing point reaches this dot in the paper, the operator knows that the motion around the diagram has been completed. The direction of motion of the tracing point must always be the same as that of the hands of a watch; motion in the opposite direction will move the wheel in the wrong direction and give a negative reading for the area. When measuring diagrams with loops, like Fig. 3, move the tracing point so that it will follow the outline of the loops in a direction opposite to the direction of motion of the hands of a watch, as is indicated by the arrow-heads on the diagram. This will cause the instrument automatically to subtract the areas of the loops from the area of the main part of the diagram.

An excellent check on the work is to start with the recording wheel at zero and pass the tracing wheel around the diagram two or three times, noting the reading of the wheel each time the tracing point returns to the point of starting. Each reading of the wheel divided by the number of times the outline

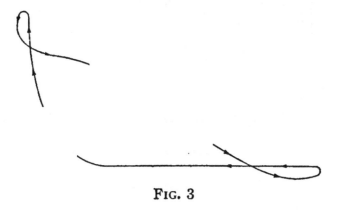

FIG. 3

of the diagram has been traced should give, very nearly, the value of the first reading; if there is considerable difference between the first reading and the value obtained by dividing the second reading by 2 or the third reading by 3, it is an indication that an error has been made, and the work should be repeated. If the difference is small, the work may be assumed to be satisfactory and the value to be used for the area or the *M. E. P.* may be taken as the average found by dividing the last reading by the number of times the tracing point passed around the diagram.

Finding **M. E. P.** by Ordinates.—The *M. E. P.* may be found from the diagram by the aid of two triangles, a scale, and a hard lead pencil; if two triangles are not available, a single triangle and a straightedge will suffice. Lines perpendicular to the atmospheric line and tangent to the two ends of the diagram must first be drawn. The perpendicular distance between these tangents will be the length of the diagram, and this length must be divided into some number of equal parts; 10 or 20 parts are the most convenient, but any other number may be used. Midway between each pair of points of division draw a line parallel to the two tangents; the part of this line included between the lines of the diagram is the middle *ordinate* of its corresponding space. The sum of the lengths of all of these middle ordinates divided by the number of ordinates is the mean ordinate and gives, approximately, the average height of the diagram. The length of the mean ordinate

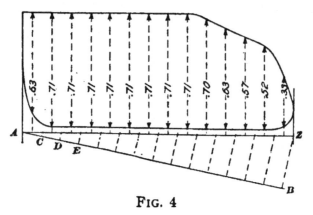

FIG. 4

should agree very nearly with the value obtained by dividing the area of the diagram, as measured by a planimeter, by the length of the diagram. The *M. E. P.* is found by multiplying the length of the mean ordinate by the scale of the spring with which the diagram was taken. A diagram thus divided into equal parts, with the lengths of the ordinates marked thereon, is shown in Fig. 4. The sum of the lengths of the ordinates is 9.06 in. As there are 14 ordinates the length of the mean ordinate is $\dfrac{9.06}{14}$ in., and if the diagram was taken

with an 80 spring, the *M. E. P.* is $\dfrac{9.06}{14} \times 80 = 51.77$ lb. per sq. in.

If a scale graduated to correspond with the scale of the

spring is available, the *M. E. P.* may be obtained by measuring the ordinates in pounds instead of in inches; the sum of the lengths of the ordinates as so measured divided by their number gives the *M. E. P.* of the diagram. For example, let the scale of the spring be 40; then each $\frac{1}{40}$ in. in the length of an ordinate represents a pressure of 1 lb. per sq. in., and by measuring the length of an ordinate with a scale graduated in fortieths of an inch, the number of pounds of pressure represented by that ordinate is found.

A convenient method of finding the sum of the lengths of the ordinates of a diagram, and one that is especially to be recommended when a decimal scale is not available, is the following: Take a strip of paper having a straight edge a little longer than the sum of the lengths of the ordinates. Lay this strip along the first ordinate. From the point on the strip representing one end of the first ordinate lay off the length of the next ordinate. In the same way lay off on the strip the length of each of the ordinates in succession. The length of the strip included between the extreme, or first and last, points so marked will be equal to the sum of the lengths of the ordinates, and this length divided by the number of ordinates will give the length of the mean ordinate.

Locating the Ordinates.—The length of the diagram will seldom be divisible into equal parts that can readily be laid off by a scale, and to divide the length into equal parts by a cut-and-try process will be found very tedious. These difficulties may, however, be overcome by an application of a simple geometrical principle, in the manner illustrated in Fig. 4. The tangent lines at the ends of the diagram are drawn perpendicular to the atmospheric line *AZ*. Suppose, now, that it is desired to have fourteen ordinates. Draw any other line from *A*, as *AB*, at a small angle to *AZ*, and then lay off any convenient distance *AC* fourteen times successively, along *AB*. Connect the last point *B* with *Z*, and from the other points *D*, *E*, etc. draw lines parallel to *BZ* until they intersect *AZ*. These points of intersection will divide the line *AZ* into fourteen equal spaces. The middle points of these spaces can then be located by direct measurement and the ordinates may be erected at these middle points.

14

Approximate **M. E. P.**—If an indicator is not available, so that diagrams may be taken in order to determine the M. E. P. of an engine, the value of the M. E. P. may be estimated by the formula

$$P = .9[C(p+14.7)-17],$$

in which P = M. E. P., in pounds per square inch;

C = constant corresponding to cut-off, taken from accompanying table;

p = boiler pressure, in pounds per square inch, gauge.

The foregoing formula applies only to a simple non-condensing engine. If the engine is a simple condensing engine, the formula should be altered by substituting for 17 the pressure existing in the condenser, in pounds per square inch.

CONSTANTS USED IN CALCULATING M. E. P.

Cut-off	Constant	Cut-off	Constant	Cut-off	Constant
$\frac{1}{6}$	.545	$\frac{3}{8}$	.773	$\frac{2}{3}$	.943
$\frac{1}{5}$	.590	.4	.794	.7	.954
$\frac{1}{4}$	.650	$\frac{1}{2}$	.864	$\frac{3}{4}$	.970
$\frac{1}{3}$	.705	.6	.916	.8	.981
$\frac{3}{8}$	.737	$\frac{5}{8}$	.927	$\frac{7}{8}$	.993

In this table, the fraction indicating the point of cut-off is obtained by dividing the distance that the piston has traveled when the steam is cut off by the whole length of the stroke; that is, it is the apparent cut-off. It is to be observed that this rule cannot be applied to a compound engine or to any other engine in which the steam is expanded in successive stages in several cylinders.

EXAMPLE.—Find the approximate M. E. P. of a non-condensing engine cutting off at $\frac{1}{2}$ stroke, if the boiler pressure is 80 lb., gauge.

SOLUTION.—According to the table, the constant corresponding to cut-off at $\frac{1}{2}$ stroke is C = .864. Then, applying the formula, $P = .9[.864(80+14.7)-17] = 58.34$ lb. per sq. in.

HORSEPOWER AND STEAM CONSUMPTION

Indicated Horsepower.—The indicator furnishes the most ready method of measuring the pressures on the piston of a steam engine and, in consequence, of determining the amount of work done in the cylinder and the corresponding horsepower. The power measured by the use of the indicator is called the *indicated horsepower.* It is the total power developed by the action of the net pressures of the steam on the two sides of the moving piston. The indicated horsepower is generally represented by the initials I. H. P.

Friction Horsepower.—The part of the indicated horsepower that is absorbed in overcoming the frictional resistances of the moving parts of the engine is termed the *friction horsepower.* If the engine is running light, or with no load, all the power developed in the cylinder is absorbed in keeping the engine in motion, and the friction horsepower is equal to the indicated horsepower. This principle furnishes a simple approximate method of finding the friction horsepower of a given engine; as, however, the friction between the surfaces increases with the pressure, the power absorbed in overcoming the engine will be greater as the load on the engine is increased.

Net Horsepower.—The difference between the indicated horsepower and the friction horsepower is the *net horsepower.* It is the power that the engine delivers through the flywheel or shaft to the belt or the machine driven by it, and is sometimes called the *delivered horsepower.* Since the power that an engine is capable of delivering when working under certain conditions is often measured by a device known as a Prony brake, the net horsepower is frequently called the *brake horsepower,* abbreviated B. H. P.

Mechanical Efficiency.—The *mechanical efficiency* of an engine is the ratio of the net horsepower to the indicated horsepower; or it is the percentage of the mechanical energy developed in the cylinder that is utilized in doing useful work. To find the efficiency of an engine, when the indicated and net horsepowers are known, divide the net horsepower by the indicated horsepower.

General Rule for Calculating I. H. P.—Knowing the dimensions and speed of the engine and the mean effective pressure on the piston, all the data for finding the rate of work done in the engine cylinder expressed in horsepower are at hand.

Let H = indicated horsepower of engine;

$P = M.$ E. P., in pounds per square inch;

A = area of piston, in square inches;

L = length of stroke, in feet;

N = number of working strokes per minute.

Then,
$$H = \frac{P\ L\ A\ N}{33,000}$$

In a double-acting engine, or one in which the steam acts alternately on both sides of the piston, the number of working strokes per minute is twice the number of revolutions per minute. For example, if a double-acting engine runs at a speed of 210 R. P. M. there are 420 working strokes per minute. A few types of engines, however, are single-acting; that is, the steam acts on only one side of the piston. Such are the Westinghouse, the Willans, and others. In this case, only one stroke per revolution does work, and, consequently, the number of strokes per minute to be used in the foregoing formula is the same as the number of revolutions per minute. Unless it is specifically stated that an engine is single-acting, it is always understood, when the dimensions of a steam engine are given, that a double-acting engine is meant.

Piston Speed.—The total distance traveled by the piston in 1 min. is called the *piston speed*. It is customary to take the stroke in inches. Then, to find the piston speed, multiply the stroke in inches by the number of strokes and divide by 12; or, letting S represent the piston speed, $S = \dfrac{l\ N}{12}$, where l is the stroke in inches. But $N = 2\ R$, where R represents the number of revolutions per minute. Hence,

$$S = \frac{lN}{12} = \frac{l \times 2\ R}{12} = \frac{l\ R}{6}$$

EXAMPLE.—An engine with a 52-in. stroke runs at a speed of 66 R. P. M. What is the piston speed?

SOLUTION.—By the formula, $S = \dfrac{52 \times 66}{6} = 572$ ft. per min.

The piston speeds used in modern practice are about as follows:

Type of Engine	Piston Speed Ft. per Min.
Small stationary engines..............	300 to 600
Large stationary engines..............	600 to 1,000
Corliss engines......................	400 to 750
Marine engines......................	200 to 1,200

Allowance for Area of Piston Rod.—It is generally considered sufficiently accurate to take the total area of one side of the piston as the area to be used in calculating the horsepower of an engine. The effective area of one side of the piston is, however, reduced by the sectional area of the piston rod, and if it is important that the power be calculated with the greatest practical degree of accuracy, an allowance for the area of the piston rod must be made. This is done by taking as the piston area one-half the sum of the areas exposed to steam pressure on the two sides of the piston. Thus, if a piston is 30 in. in diameter with a 6-in. piston rod, the average area is

$$\frac{30^2 \times .7854 + (30^2 \times .7854 - 6^2 \times .7854)}{2} = 692.72 \text{ sq. in.}$$ If the

piston rod is continued past the piston so as to pass through the head-end cylinder head, that is, if the piston has a *tailrod*, allowance must be made for the tailrod. Thus, with a piston 30 in. in diameter, a piston rod 6 in. in diameter, and a tailrod 5 in. in diameter, the average area is

$$\frac{(30^2 \times .7854 - 5^2 \times .7854) + (30^2 \times .7854 - 6^2 \times .7854)}{2} = 682.9 \text{ sq. in.}$$

Stating Sizes of Engines.—The size of a simple engine, that is, an engine having but one cylinder, is commonly stated by giving the diameter of the cylinder, followed by the length of the stroke, both in inches. Thus, a simple engine having a cylinder 12 in. in diameter and a stroke of 24 in. would be referred to as a 12″×24″ engine, the multiplication sign in this case serving merely to separate the two numbers. The sizes of compound and multiple-expansion engines are designated in a similar fashion. Thus, a compound engine with a high-pressure cylinder 11 in. in diameter, a low-pressure cylinder 20 in. in diameter, and a stroke of 15 in. would be referred to as an $11''$ and $20'' \times 15''$ compound engine. In the same way, a $14''$, $22''$, and $34'' \times 18''$ triple-expansion engine would be one in which the diameters of the cylinders are 14 in., 22 in., and 34 in., and the stroke is 18 in.

Cylinder Ratios.—The cylinders of compound and multiple-expansion engines increase in diameter from the high-pressure to the low-pressure end, and it is customary to refer to their relative sizes by means of cylinder ratios. As all the cylinders have the same length of stroke, the volumes of the several cylinders are in proportion to the areas of the cylinders, and therefore in proportion to the squares of the diameters. The area of the high-pressure cylinder is taken as unity, and the other areas are referred to it, and the ratios of these areas, or the ratios of the squares of the diameters, are called the *cylinder ratios*. For example, a triple-expansion engine having cylinders 12 in., 20 in., and 34 in., in diameter will have the cylinder ratios of $12^2 : 20^2 : 34^2$, or $144 : 400 : 1{,}156$, which reduces to $1 : 2.78 : 8.03$; that is, the intermediate cylinder is 2.78 times as large as the high-pressure cylinder, and the low-pressure cylinder is 8.03 times as large as the high-pressure cylinder. If there are two cylinders to one stage of expansion, as, for example, two low-pressure cylinders, the sum of their areas must be used in finding the cylinder ratios. Thus, if there had been two 24-in. low-pressure cylinders instead of one 34-in. cylinder, in the foregoing case, the cylinder ratios would have been $12^2 : 20^2 : 2 \times 24^2$, or $144 : 400 : 1{,}152$, which reduces to $1 : 2.78 : 8$.

Horsepower of Compound Engines.—The indicated horsepower of a compound or triple-expansion engine may be calculated from the indicator diagrams in exactly the same manner as with any simple engine, considering each cylinder as a simple engine and adding the horsepowers of the several cylinders together. In taking the indicator cards from a compound engine, the precaution of taking the cards simultaneously from all cylinders must be observed, especially when the engine runs under a variable load, because, otherwise, an entirely wrong distribution of power may be shown, and there may also be a great variation between the indicated horsepower really existing and that calculated from diagrams taken at different times.

Referred Mean Effective Pressure.—The indicated horsepower of compound engines is sometimes found by referring the mean effective pressure of the high-pressure cylinder to the low-pressure cylinder and calculating the horsepower of the engine on the assumption that all the work is done in the low-pressure cylinder. To do this, the mean effective pressures of the two cylinders are found from indicator diagrams; the mean effective pressure of the high-pressure cylinder is then divided by the ratio of the volume of the low-pressure cylinder to that of the high-pressure cylinder; and the quotient is added to the mean effective pressure of the low-pressure cylinder, the sum being the *referred mean effective pressure*. This sum is then taken as the mean effective pressure of the engine, and the area of the low-pressure piston as the piston area; with these data, the length of stroke and the number of strokes, the horsepower is computed as for any simple engine. In the case of a triple-expansion engine, the mean effective pressures of the high-pressure and intermediate cylinders are referred to the low-pressure cylinder and added to its mean effective pressure. Thus, suppose that in a 12″, 20″, and 34″×30″ engine the mean effective pressures are 83.2 lb., 27.8 lb., and 10.6 lb., respectively. Then, the referred *M. E. P.* is $\dfrac{83.2}{8.03}$

$+\dfrac{27.8}{2.78}+\dfrac{10.6}{1}=10.4+10+10.6=31$ lb., and this value must be

substituted for P on finding the horsepower of the engine. While this method shortens the labor of computing the horsepower, it obviously does not show the distribution of work between the cylinders.

Dynamometers.—Dynamometers are instruments for measuring power. They are divided into two main classes: *absorption dynamometers* and *transmission dynamometers*. The most common form of absorption dynamometer is the *Prony brake*, which consists simply of a friction brake designed to absorb in friction and measure the work done by a motor, or the power given out by a shaft. A transmission dynamometer is used to measure the power required to drive a machine or do other work; thus, to determine the power required to run the shafting in a mill, a transmission dynamometer would be interposed between the shafting and the source of power, and by suitable belt connections the shafting would be driven through the dynamometer, from which the power could be determined. As transmission dynamometers do not enter into the work of the steam engineer, they will not be treated of here.

Prony Brake.—The brake horsepower of steam engines is commonly determined by means of some form of friction brake. One construction of Prony brake is shown in Fig. 1. It consists of two wooden blocks *a* and *b* formed so as to fit

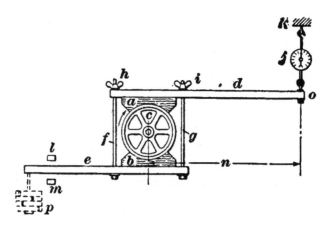

FIG. 1

the face of the iron pulley *c* on the shaft of the engine to be tested. To the blocks are fixed the two long arms *d* and *e*, and the whole is clamped together by means of the bolts *f* and *g* and the nuts *h* and *i*. By tightening these nuts the blocks *a* and *b* may be pressed more tightly against the face of the pulley, thus increasing the friction at the surface of the pulley. If the engine rotates so as to turn the pulley in the direction indicated by the arrow, the friction on the face of the pulley will tend to drag the blocks around in the same direction; but as the end of the arm *d* carries a spring balance *j* attached to a stationary support *k*, the tendency to turn is indicated by a pull on the spring balance. The tighter the

nuts *h* and *i* are screwed up, the greater will be the pull exerted
on the spring balance.   The arm *e* is extended and fits between
two stops *l* and *n*, so as to prevent excessive movement of the
brake.   While a test is being made, the arm *e* should not
touch either stop.   This can easily be attained by regulating
the tightness of the nuts *h* and *i*.   The stops, however, should
never be dispensed with, as a sudden reversal of the engine or
an unexpected gripping of the pulley by the wooden blocks
might swing the brake arms around and injure the workmen.
The distance *n* from the center of the pulley *c* to a plumb-
line suspended from the center of the bolt *o* should be measured
very accurately, with the arm *d* in a level position.   This

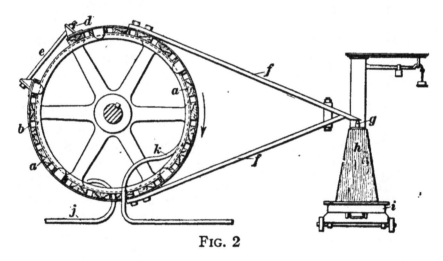

Fig. 2

distance *n* is used in determining the brake horsepower, as will
be explained later.

Another form of friction brake is illustrated in Fig. 2.   In
this type, there are a number of wooden blocks *a* fastened to an
iron band *b* and surrounding the pulley *c*.   The friction
between the blocks and the pulley may be altered by loosening
or tightening the nut *d* on the bolt *e* that joins the ends of
the band *b*.   To the bands are fastened the arms *f*, which are
bolted together at their outer ends and rest on a knife edge *g*
fastened to a block *h* resting on a platform scales *i*.   When the
pulley is rotated in the direction indicated by the arrow, the
outer ends of the arms *f* press down on the block *h* and

the pressure may be measured by the scales. The distance between the center of the shaft to which the pulley is keyed and the knife edge *g* should be measured carefully, as it represents the effective length of the brake arm.

Rope Brake.—A form of brake requiring less space than the Prony brake is the rope brake, shown in Fig. 3. It is used in determining the horsepower of engines of moderate size. The pulley *a* on the engine shaft is encircled by a double rope *b* to which are fixed blocks *c* that bear against the face of the

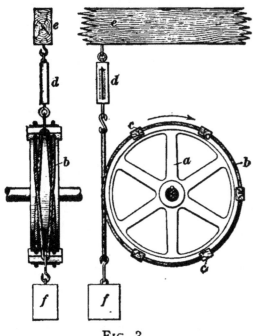

pulley. One end of the rope is attached to a spring balance *d* suspended from a beam *e*, and the other end carries a weight *f* that may be varied. The friction of the brake on the pulley is increased by adding to the weight *f*. The direction of rotation is indicated by the arrow. The brake arm in this type of brake is the distance from the center of the pulley to the center of the rope, and this distance should be measured very carefully and quite accurately, as it is required in calculating the horsepower.

Fig. 3

Cooling of Brake Pulleys.—It is essential that the pulleys of brakes should be well lubricated and for all except small powers, means must be provided for conducting away the heat generated by friction. If there are internal flanges on the brake wheel, water can be run on the inside of the rim, the flanges serving to retain the water at the sides and centrifugal force to keep it in contact with the rim. A funnel-shaped scoop can be used to remove the water. It should be attached to a pipe and placed so as to scoop out the water, which should flow continuously. An arrangement of this kind is shown in Fig. 2, in which water is led into the rim of the pulley by the pipe *j* and is scooped out and led away by the pipe *k*.

Data for Brake Horsepower.—It is necessary to know three factors in order to determine the brake horsepower. These are (1) the *net* pull or pressure exerted at the end of the brake arm, (2) the length of the brake arm, and (3) the number of revolutions per minute of the brake pulley. The work done by the engine is converted into heat by the friction between the wooden blocks and the face of the pulley, and the resistance offered by the brake to the rotation of the pulley is a measure of the work done. This resistance cannot always be measured conveniently at the surface of the pulley, but it can be measured at the end of the brake arm by the scales or the spring balance.

Net Pull or Pressure.—With the brake shown in Fig. 1, there will be a pull in the spring balance due to the weight of the arm *d*, when the nuts *h* and *i* are slacked and the brake is loose on the pulley. This pull should be observed carefully, and should be subtracted from the pull registered by the spring balance during a test. The difference will be the *net pull*, from which the brake horsepower is calculated. If the arm *d* is counterbalanced accurately by adding weights to the arm *e*, as shown dotted at *p*, it will be unnecessary to perform this subtraction. The net pull will then be indicated directly by the reading of the spring balance.

The net pressure must also be calculated if the brake shown in Fig. 2 is used. The nut *d* should be loosened, and the arms *f* should be worked up and down, to make sure that the blocks are loose on the pulley. The ends of the arms should then be rested on the knife edge *g* and the reading of the scale beam should be observed. This weight represents the pressure due to the block *h* and the unbalanced arms *f*, and it must be subtracted from the reading of the scale beam observed during a test in order to determine the net pressure.

In the case of the rope brake shown in Fig. 3, the net pull is easily determined by subtracting the reading indicated by the spring balance from the weight, in pounds, applied at *f*.

Calculation of Brake Horsepower.—When the necessary data, as previously noted, have been determined, the brake horsepower may be calculated by the formula

$$H_b = \frac{2 \pi R W N}{33,000},$$

in which H_b = brake horsepower;

π = 3.1416;

R = length of brake arm, in feet;

W = net pull or pressure, in pounds;

N = number of revolutions per minute.

EXAMPLE.—A Prony brake with an arm 6 ft. long was placed on the flywheel of an engine running at 200 R. P. *M*. What brake horsepower was being developed when the net pressure was 140 lb.?

SOLUTION.—Applying the formula,

$$H_b = \frac{2 \times 3.1416 \times 6 \times 140 \times 200}{33,000} = 32 \text{ H. P.}$$

CONDENSERS

Types of Condensers.—There are two types of condensers in general use, namely, the *surface condenser* and the *jet condenser*. In the former, the exhaust steam comes in contact with a large area of metallic surface that is kept cool by contact with cold water. In the latter, the exhaust steam, on entering the condenser, comes in contact with a jet of cold water. In either case, the entering steam is condensed to water, and in consequence a partial vacuum is formed. If enough cold water were used, the steam on entering would instantly condense and a practically perfect vacuum would be obtained were it not for the fact that the feedwater of the boiler always contains a small quantity of air, which passes with the exhaust steam into the condenser and therefore partly destroys the vacuum. To get rid of this air, the condenser is fitted with an air pump, which pumps out both the air and the water formed by condensation.

Surface Condensers.—In the surface condenser, the exhaust steam and the injection water are kept separate throughout their course through the condenser; and the condensed steam leaves the condenser as fresh water, free from the impurities contained in the injection water. The water of condensation from a surface condenser is therefore fit to be used as boiler feed, except that it contains oil used for cylinder lubrication, which can be eliminated by means of an oil separator, regardless of the quality of the water used to condense it. It is for this reason that the surface condenser, in spite of its greater complication, cost, size, and weight, as compared with the jet condenser, is used instead of the latter where the supply of injection water is unfit for use as boiler feed. Thus, the surface condenser is used altogether in marine work, except for vessels navigating clean fresh water like that of the Great Lakes, in order to avoid the use of sea-water in the boilers.

In the surface condenser the steam may be outside and the water inside the tubes, or the reverse. If the water is inside the tubes, it should enter at the bottom of the condenser and be discharged at the top. This brings the coldest water into contact with the partly condensed steam, and the warmest water into contact with the hot entering steam. When the water is outside the tubes, it is necessary to fit baffle plates on the water side to force the water into a definite and regular circulation, and to prevent it from going directly from inlet to outlet and also to prevent the water from arranging itself in layers according to temperature, with the coldest water on the bottom and the hottest water on top. The outlet should be well above the top-row of tubes. A solid body of water above the top row of tubes is thus assured, and the accumulation of a stagnant body of hot water in the top of the condenser is prevented by its being continually drawn off by the circulating pump and replaced by cooler water from beneath.

Air tends to accumulate in the top of the water side of a surface condenser. This is particularly inconvenient where the water is inside the tubes, as the air fills the top rows of tubes and excludes the water, destroying their value as cooling surfaces. To prevent this, an air valve must be provided, as high up on the water side as possible, in all surface condensers, by which the air can be drawn off whenever it becomes troublesome. Drain valves and pipes should be provided at the bottom.

As the condensed steam from the surface condenser is generally pumped back into the boiler as feedwater, it is desirable to have it as hot as possible; but it must be remembered that it is impossible to get the feedwater from the condenser at a higher temperature than that of saturated steam at the absolute pressure existing in the condenser.

It will be considerably cooler than this if, after being condensed, it is allowed to lie in the bottom of the condenser and give up its heat to the circulating water. The heat thus given up is a total loss, and should be avoided by connecting the air-pump suction to the lowest point of the condenser and by shaping the bottom of the condenser so that the water will drain rapidly into the air-pump suction.

Cooling Water **for** Surface Condenser.—The amount of cooling water required in the case of a surface condenser may be found by the formula

$$Q = \frac{H - (t - 32)}{t_2 - t_1},$$

in which Q = number of pounds of cooling water required to condense 1 lb. of steam;

H = total heat above 32° of 1 lb. of steam at pressure at release;

t = temperature of condensed steam on leaving condenser;

t_1 = temperature of cooling water on entering condenser;

t_2 = temperature of cooling water on leaving condenser.

EXAMPLE.—Steam exhausts into a surface condenser from an engine cylinder at a pressure of 6 lb., absolute; the temperature of the condensing water on entering is 55° F., and on leaving it is 100° F.; the temperature of the condensed steam on leaving the condenser is 125° F. How many pounds of cooling water are required per pound of steam?

SOLUTION.—The total of 1 lb. of steam at 6 lb., absolute, from the Steam Table, is 1,133.8 B. T. U. Then, substituting the values of H, t, t_1, and t_2 in the formula,

$$Q = \frac{1,133.8 - (125 - 32)}{100 - 55} = \frac{1,040.8}{45} = 23.13 \text{ lb.}$$

Injection Water for Jet Condenser.—The quantity of injection water required for a jet condenser may be found by the formula

$$Q = \frac{H - (t - 32)}{t - t_1},$$

in which Q = number of pounds of injection water required to condense 1 lb. of steam;

H = total heat above 32° of 1 lb. of steam at pressure at release;

t = temperature of mixture of injection water and condensed steam on leaving the condenser;

t_1 = temperature of injection water on entering the condenser.

EXAMPLE.—Steam is exhausted into a jet condenser from an engine cylinder at a pressure of 10 lb., absolute; the temperature of the injection water on entering is 60° F., and on leaving 140° F. How much injection water is required per pound of steam?

SOLUTION.—The total heat above 32° of 1 lb. of steam at 10 lb. absolute, from the Steam Table, is 1,140.9 B. T. U. Then, substituting the values of H, t, and t_1 in the formula,

$$Q = \frac{1,140.9 - (140 - 32)}{140 - 60} = \frac{1,032.9}{80} = 12.91 \text{ lb.}$$

ENGINE MANAGEMENT

STARTING AND STOPPING

Warming Up.—About 15 or 20 min. before starting the engine, the stop-valves should be raised just off their seats and a little steam should be allowed to flow into the steam pipe. The drain cock on the steam pipe just above the throttle should be opened. When the steam pipe is thoroughly warmed up and steam blows through the drain pipe, the drain cock should be closed and the throttle opened just enough to let a little steam flow into the valve chest and cylinder; or if a by-pass around the throttle is fitted, it may be used. The cylinder relief valves, or drain cocks, and also the drain cocks on the valve chest and the exhaust pipe should be opened, if the engine is non-condensing. If the cylinders are jacketed, steam should be turned into the jackets and the jacket drain cocks should be opened. While the engine is warming up, the oil cups and the sight-feed lubricator may be filled. A little oil may be put into all the small joints and journals that are not fitted with oil cups. The guides should be wiped off with oily waste and oiled. By this time the engine is getting warm. If the cylinder is fitted with by-pass valves, they should be used to admit steam to both ends of the cylinder. In general, all cylinders, especially if they are large and intricate castings, should be warmed up slowly, as sudden and violent heating of a cylinder of this character is very liable to crack the casting by unequal expansion.

An excellent and economical plan for warming up the steam pipe and the engine is to open the stop-valves and throttle valve at the time or soon after the fires are lighted in the boilers, permitting the heated air from the boilers to circulate through the engine, thus warming it up gradually and avoiding the accumulation of a large quantity of water of condensation in the steam pipe and cylinder. When pressure shows on the boiler gauge or steam at the drain pipes of the engine, the stop-valves and throttle may be closed temporarily, but not hard down on their seats. When this method of warming up the engine is adopted, the safety valves should not be opened while steam is being raised.

Danger of Water Hammer.—Stop-valves and throttle valves should never be opened quickly or suddenly and thus permit a large volume of steam to flow into a cold steam pipe or cylinder. If this is done, the first steam that enters will be condensed and a partial vacuum will be formed. This will be closely followed by another rush of steam with similar results, and so on until a mass of water will collect, which will rush through the steam pipe and strike the first obstruction, generally the bend in the steam pipe near the cylinder, with great force, and in all probability will carry it away and cause a disaster. This is called *water hammer* and has caused many serious accidents. Before turning steam into any pipe line or into a cylinder, all drain valves should be opened.

Easing of Throttle Valve.—Another precaution that should be taken is the easing of the throttle valve on its seat before steam is let into the main steam pipe; otherwise, the unequal expansion of the valve casing may cause the valve to stick fast and thereby give much trouble. Even if a by-pass pipe is fitted around the throttle, it would be better not to depend on it. Considerable space has been devoted to the subject of warming up and draining the water out of the steam pipe and engine on account of its importance. Water being noncompressible, it is an easy matter to blow off a cylinder head or break a piston if the engine is started when there is a quantity of water in the cylinder.

Oil and Grease Cups.—The last thing for the engineer to do before taking his place at the throttle preparatory to starting the engine, provided he has no oiler, is to start the oil and grease cups feeding. It is well to feed the oil liberally at first, but not to the extent of wasting it; finer adjustment of the oiling gear can be made after the engine has been running a short time and the journals are well lubricated.

Starting Non-Condensing Slide-Valve Engine.—A noncondensing slide-valve engine is started by simply opening the throttle; this should be done quickly in order to jump the crank over the first dead center, after which the momentum of the flywheel will carry it over the other centers. The engine should be run slowly at first, gradually increasing the revolutions to the normal speed. When the engine has reached full speed, the drain pipes should be examined; if dry steam is blowing through them, the drain cocks should be closed. If water is being delivered, the drain cocks should remain open until steam blows through and should then be closed.

Stopping Non-Condensing Slide-Valve Engine.—To stop a non-condensing slide-valve engine, it is only necessary to shut off the supply of steam by closing the throttle, but care should be taken not to let the engine stop on the dead center. After the engine is stopped, the oil feed should be shut off and the main stop-valve closed. The valve should be seated, but without being jammed hard down on its seat. The drain cocks on the steam pipe and engine may or may not

15

be opened, according to circumstances. It will do no harm
to allow the steam to condense inside the engine, as the engine
will then cool down more gradually, which lessens the danger
of cracking the cylinder casting by unequal contraction. All
the water of condensation should be drained from the engine
before steam is again admitted to it.

Starting Condensing Slide-Valve Engine.—Before the main
engine is started, the air pump and circulating pump should
be put into operation and a vacuum formed in the condenser;
this will materially assist the main engine in starting promptly.
Prior to starting the air and circulating pumps, the injection
valve should be opened to admit the condensing water into
the circulating pump; the delivery valve should also be opened
at this time. If an ordinary jet condenser is used, no ciren-
lating pump is required, the water being forced into the condenser by the pressure of the atmosphere. If the air pump is
operated by the main engine, a vacuum will not be formed
in the condenser until after the engine is started and at least
one upward stroke of the air pump is made. In this case
the injection valve must be opened at the same moment the
engine is started; otherwise the condenser will get hot and a
mixture of air and steam accumulate in it and prevent the
injection water from entering. When this occurs, it is necessary to pump cold water into the condenser by one of the
auxiliary pumps through a pipe usually fitted for that purpose;
if such a pipe has not been provided, it may be found necessary to cool the condenser by playing cold water on it through
a hose.

Stopping Condensing Slide-Valve Engine.—The operation
of stopping a slide-valve surface-condensing engine is precisely
similar to that of stopping a non-condensing engine of the same
type, with the addition that after the main engine is stopped
the air and circulating pumps are also stopped, and in the
same way, that is, by closing the throttle, after which the
injection valve and the discharge valve should be closed and
the drain cocks opened. With a jet condenser, the operation
of stopping the engine is the same as the above, with the
exception that the injection valve should be closed at the same
moment that the engine is stopped.

Starting Simple Corliss Engine.—In the Corliss engine, the eccentric rod is so constructed and arranged that it may be hooked on or unhooked from the eccentric pin on the wrist-plate at the will of the engineer. After all the preliminary operations have been attended to, the starting bar is shipped into its socket in the wristplate and the throttle is opened. The starting bar is then vibrated back and forth by hand, by which the steam and exhaust valves are operated through the wristplate and valve rods; as soon as the cylinder takes steam, the engine will start. After working the starting bar until the engine has made several revolutions and the flywheel has acquired sufficient momentum to carry the crank over the dead centers, the hook of the eccentric rod should be allowed to drop upon the pin on the wristplate. As soon as the hook engages with the pin, the starting bar is unshipped and placed in its socket in the floor. The way to determine in which direction the starting bar should be first moved to start the engine ahead is to note the position of the crank, from which the direction in which the piston is to move may be learned. This will indicate which steam valve to open first; it will then be an easy matter to determine in which direction the starting bar should be moved. If the engine is of the condensing type, the same course of procedure in starting the air and circulating pumps should be followed as with the simple condensing slide-valve engine.

Stopping Simple Corliss Engine.—A Corliss engine is stopped by closing the throttle and unhooking the eccentric rod from the pin on the wristplate; this is done by means of the unhooking gear provided for the purpose. As soon as the eccentric rod is unhooked from the pin, the starting bar is shipped into its socket in the wristplate and the engine is worked by hand to any point in the revolution of the crank at which it is desired to stop the engine. The procedure is then the same as for the simple slide-valve engine. After stopping a Corliss condensing engine, the same course should be followed as with a slide-valve condensing engine in regard to draining cylinders, closing stop-valves, etc.

Starting Compound Slide-Valve Engine.—Before starting a compound engine, the high-pressure cylinder is warmed up in the same manner as a simple engine. To get the steam into the low-pressure cylinder is, however, an operation that will depend on circumstances. If the cylinders are provided with pass-over valves, it will be necessary only to open them to admit steam into the receiver and from thence into the low-pressure cylinder. If the cylinders are not fitted with pass-over valves, the steam can usually be worked into the receiver and low-pressure cylinder by operating the high-pressure valves by hand. Sometimes compound engines are fitted with starting valves, which greatly facilitate the operations of warming up and starting. Usually a compound engine will start upon opening the throttle.

If the high-pressure crank of a cross-compound engine is on its center and the low-pressure engine will not pull it off, it must be jacked off. If the pressure of steam in the receiver is too high, causing too much back pressure in the high-pressure cylinder, the excess of pressure must be blown off through the receiver safety valve; if the pressure in the receiver is too low to start the low-pressure piston, more steam must be admitted into the receiver. If the engine is stuck fast from gummy oil or rusty cylinders, all wearing surfaces must be well oiled and the engine jacked over at least one entire revolution. If the cut-offs are run up, they should be run down, full open. If there is water in the cylinders, it should be blown out through the cylinder relief or drain valves, and if there is any obstruction to the engine turning, it should be removed.

If the crank of a tandem compound engine is on the center, it must be pulled or jacked off. If the high-pressure crank of a cross-compound engine is on the center, it may or may not be possible to start the engine by the aid of the low-pressure cylinder, depending on the valve gear and the crank arrangement. When the cranks are 180° apart, which is a very rare arrangement, the crank must be pulled or jacked off the center. When the cranks are 90° apart and a pass-over valve is fitted, live steam may be admitted into the receiver and thence into the low-pressure cylinder, in order to start the engine. When no pass-over is fitted, but the engine has a link motion, sufficient steam to pull the high-pressure crank off the center can generally be worked into the low-pressure cylinder by working the links back and forth.   When no pass-over is fitted, but the high-pressure engine can have its valve or valves worked by hand, steam can be got into the low-pressure engine by working the high-pressure valve or valves back and forth by hand. If no way exists of getting steam into the low-pressure cylinder while the high-pressure crank is on a dead center, it must be pulled or jacked off.

If the air and circulating pumps are attached to and operated by the main engine, a vacuum cannot be generated in the condenser until after the main engine has been started.   Consequently, in this case, there is no vacuum to help start the engine; therefore, if it is tardy or refuses to start, it will be necessary to resort to the jacking gear and jack the engine into a position from which it will start.   A vacuum having been generated in the condenser beforehand, the pressure in the receiver acting on the low-pressure piston causes the engine to start promptly, even though the high-pressure crank may be on its center.

Stopping Compound Slide-Valve Engine.—Compound slide-valve engines, whether condensing or non-condensing, are stopped by closing the throttle, and, if a reversing engine, throwing the valve gear into mid-position.   If the stop is a permanent one, the usual practice of draining the engine, steam chests, and receiver, closing stop-valves, stopping the oil feed, etc. should be followed.   If the engine is intended to run in both directions in answer to signals, as in the cases of hoisting, rolling-mill, and marine engines, the operator, after stopping the engine to signal, should immediately open the throttle very slightly, in order to keep the engine warm, and stand by for the next signal.   If the engine is fitted with an independent or adjustable cut-off gear, it should be thrown off, that is, set for the greatest cut-off, for the reason that the engine may have stopped in a position in which the cut-off valves in their early cut-off positions would permit little or no steam to enter the cylinders, in which case the engine will not start promptly, and perhaps not at all.   While waiting for the signal, the cylinder drain valves should be opened and any water that may be in the cylinders should be blown out.   When dry steam blows through the drains, the cylinders are clear of water.

When the signal to start the engine is received, it is only necessary to throw the valve gear into the go-ahead or backing position, as the signal requires, and to operate the throttle according to the necessities of the case, for which no rule can be laid down beforehand, as the position of the throttle will depend on the load on the engine at the time.

Starting and Stopping Compound Corliss Engine.—The operation of starting and stopping a compound Corliss engine is precisely similar to that of starting and stopping a simple Corliss engine. The high-pressure valve gear only is worked by hand in starting, the low-pressure eccentric hook having been hooked on previously. The low-pressure valve gear is worked by hand only while warming up the low-pressure cylinder. The directions given for operating the simple condensing engine apply to the condensing Corliss engine, so far as the treatment of the air pump, circulating pump, and condenser is concerned.

POUNDING OF ENGINES

Loose Brasses.—Loose journal brasses are the most frequent cause of pounding in engines. The remedy for pounding of this nature is obvious. The engine should be stopped and the brasses set up gradually until the pounding ceases. In the case of shaft journals, they may be set up without stopping the engine, provided the engineer can reach them without danger of being caught in the machinery.

Brass-Bound Bearings.—It may so happen that the boxes or brasses are worn down until the edges of the upper half and those of the lower half are in contact and cannot be set up on the journal any farther; they are then said to be *brass and brass*, or *brass-bound*. In a case of this kind, the journal must be *stripped*, as it is called, when the cap and brasses are removed from a journal. The edges of the brasses are then chipped or filed off, in order to allow them to be closed in.

Liners.—It is a most excellent plan in practice to reduce the halves of the brasses so that they will stand off from each other when in place for a distance of $\frac{1}{8}$ to $\frac{3}{16}$ in. and to fill this space with hard sheet-brass liners from 20 to 22 Birmingham wire gauge in thickness, or even thinner. Should the journal become brass-bound, the cap may be slacked off and a pair of the liners slipped out without the necessity of stripping the journal.

In some instances journal-boxes are fitted with *keepers*, or *chipping pieces*, as they are sometimes called. These usually consist of cast-brass liners from $\frac{1}{4}$ to $\frac{1}{2}$ in. in thickness, having ribs or ridges cast on one side, for convenience of chipping and filing. These keepers are sometimes made of hardwood and are capable of being compressed slightly by the pressure exerted upon them during the setting-up process. When the boxes are babbitted, the body of the box is occasionally made of cast iron, in which case iron liners and keepers are used instead of brass ones.

Loose Thrust Block.—In engines fitted with some types of friction couplings, there is a thrust exerted upon the shaft in the direction of its length. This will necessitate having a *thrust bearing*, or *thrust block*, as it is sometimes called. There are a number of types of thrust bearings, but the most common is the collar thrust, which consists of a series of collars on the shaft that fit in corresponding depressions in the bearing. If these collars do not fit in the depressions rather snugly, the shaft will have end play and there probably will be more or less pounding or backlash at every change of load on the engine. This can be remedied only by putting in a new thrust bearing and making a better fit with the shaft collars, unless the rings in the bearing are adjustable, in which case the end play may be taken up by adjusting the rings.

Water in Cylinders.—Pounding often occurs in the cylinders and is frequently caused by water due to condensation or carried over from the boilers. This may be a warning that priming is likely to occur in the boilers or has already commenced. If the cylinders are not fitted with automatic relief valves, the drain cocks should be opened as quickly as possible and the throttle closed a little to check the priming.

Loose Piston.—Another source of pounding in the cylinder is a piston loose on the rod; this will result if the piston-rod nut or key backs off or the riveting becomes loose, permitting the piston to play back and forth on the piston rod. If due to backing off of the nut, the engine should be shut down instantly. There is generally very little room to spare between the piston-rod nut and the cylinder head; therefore, it cannot back off very far before it will strike and break the cylinder head. After the engine is stopped and the main stop-valve is closed, the cylinder head should be taken off and the piston nut set up as tightly as possible. As a measure of safety, a taper split pin should in all cases be fitted through the piston rod behind the nut or a setscrew should be fitted through the nut.

Slack Follower Plate.—A slack follower plate or junk ring will cause pounding in the cylinder. It seldom happens, however, that *all* the follower bolts back out at one time, but it is not an infrequent occurrence that one of the follower bolts works itself out altogether. This is a very dangerous condition of affairs, especially in a horizontal engine. If the bolts should get end on between the piston and cylinder head, either the piston or the cylinder head is bound to be broken. Therefore, if there is any intimation that a follower bolt is adrift in the cylinder, the correct procedure is to shut down the engine instantly, take off the cylinder head, remove the old bolt, and put in one having a tighter fit.

Broken Piston Packing.—Broken packing rings and broken piston springs will cause noise in the cylinder, but it is more of a rattling than a pounding, and the sound will easily be recognized by the practiced ear. There is not so much danger of a breakdown from these causes as may be supposed, from the fact that the broken pieces are confined within the space between the follower plate and the piston flange.

Piston Striking Heads.—Pounding in the cylinders of old engines is often produced by the striking of the piston against one or the other cylinder head. One of the causes of this is the wearing away of the connecting-rod brasses. Keying up the brasses from time to time has the effect of lengthening or shortening the connecting-rod, depending on the design, and this change in length destroys the clearance at one end of the cylinder by an equal amount. The remedy is to restore the rod to its original length by placing sheet-metal liners behind the brasses; this obviously will move the piston back or ahead and restore the clearance. A rather rare case of the piston striking the cylinder head is due to unscrewing of the piston rod from the crosshead, in case it is fastened by a thread and check-nut. To obviate any danger, the check-nut should be tried frequently.

Improper Steam Distribution.—The primary cause of another source of pounding is the improper setting of the steam valve, or possibly its improper design. In the case of improper setting of the valve, insufficient compression, insufficient lead, cut-off too early, and late release may all cause pounding on the centers.

Reversal of Pressure.—The effect of a reversal of pressure is clearly shown in the accompanying illustration. With the crankpin at *a* and the engine running in the direction indicated by the arrow, the connecting-rod is subjected to a pull, but after the crankpin has passed the dead center *c*, the connecting-rod is subjected to a push, in which case the rear brass, as shown at *b*, bears against the crankpin, while in the former case, 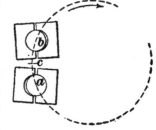 as shown at *a*, the front brass bears against the crankpin. By giving a sufficient amount of compression, the lost motion in the pins and journals is transferred gently from one side to the other before the crankpin reaches the dead center. If the compression is insufficient, there will be pounding.

Insufficient Lead.—Insufficient lead causes an engine to pound because the piston has then little or no cushion to impinge on as it approaches the end of its stroke, and it is brought to rest with a jerk. A similar effect will be produced by a late release; the pressure is retained too long on the driving side of the piston. The ideal condition is that the pressures shall be equal on both sides of the piston at a point in its travel just in advance of the opening of the steam port. The position of this point varies with the speed of the piston and other conditions that only the indicator card can reveal.

Pounding at Crosshead.—The crosshead is a source of pounding from various causes, of which the loosening of the piston rod is one of the most common. There are several methods of attaching the piston rod to the crosshead. The rod may pass through the crosshead with a shoulder or a taper, or both, on one side of the crosshead and a nut on the other; or the rod may be secured to the crosshead by a cotter, instead of the nut; or the end of the rod may be threaded and screwed into the crosshead, having a check-nut to hold the rod in place. In the case first mentioned, the nut may work loose, which would cause the crosshead to receive a violent blow, first, by the nut on one side and then by the shoulder or taper on the other at each change of motion of the piston. The remedy is to set up the nut. A similar effect will be produced if the cotter should work loose and back out. In case the piston rod is screwed into the crosshead and the rod slacks back, the danger is that the piston will strike the rear cylinder head. The check-nut should be closely watched. Pounding at the crosshead may be due to loose wristpin brasses, in which case they should be set up, but not too tightly. In case a crosshead works between parallel guides, pounding may be caused if the crosshead is too loose between the guides, and the crosshead shoes should therefore be set out.

If pounding results from the wearing down of the shoe of a slipper crosshead, a liner should be put between the shoe and the foot of the crosshead or the shoe should be set out by the adjustment provided.

Pounding in Air Pump.—Pounding in the air pump is generally produced by the slamming of the valves, caused by an undue amount of water in the pump, which will usually relieve itself after a few strokes. The pump piston, however, may be loose on the piston rod or the piston rod may be loose in the crosshead. A broken valve may also cause pounding in the air pump, all of which must be repaired as soon as detected.

Pounding in Circulating Pump.—In a circulating pump of the reciprocating type, pounding may be caused by admitting too little injection water, and the pounding may be stopped by adjusting the injection valve to admit just the right quantity. It may so happen, however, that the injection water is very cold, and to admit enough of it to stop the pounding in the circulating pump will make the feedwater too cold. To meet this contingency, an air check-valve is often fitted to the circulating pump to admit air into the barrel of the pump as a cushion for the piston; this check-valve may be kept closed when not needed to admit air. A broken valve, a piston loose on its piston rod, and a piston rod loose in the crosshead will all cause pounding in the circulating pump; they should be treated in the same manner as was specified for similar troubles in the air pump.

HOT BEARINGS

Mixtures for Reducing Friction.—Should any of the bearings show an inclination to heat to an uncomfortable point when felt by the hand, the oil feed should be increased. If the bearing continues to get hotter, some flake graphite should be mixed with the oil and the mixture should be fed into the bearing through the oil holes, between the brasses, or wherever else it can be forced in. A little aqua ammonia introduced into a hot bearing will sometimes cheek heating by converting the oil into soap by saponification, soap being an excellent lubricant. Mineral oils will not saponify.

Danger of Increasing Heating.—If, after trying the remedies just mentioned, the bearing continues to grow hotter, to the extent, for instance, of scorching the hand or burning the oil, it indicates that the brasses have been expanded by the heat and that they are gripping the journal harder and harder the hotter they get. At this stage, if the engine is not stopped or if the heating is not checked, the condition of the bearing will continue to grow worse as long as the engine is running, and may become so bad as to slow down and eventually stop the engine by excessive friction. By this time the brasses and journal will be badly cut and in bad condition generally, and the engine must be laid up for repairs.

Remedies for Increasing Heat.—The state of affairs just mentioned should not be permitted to be reached. After the simple remedies previously given have been tried and failed to produce the desired results, the engine should be stopped and the cap or key of the hot bearing should be slacked back and the engine allowed to stand until the bearing has cooled off. If necessity requires it, the cooling may be hastened by pouring cold water on the bearing, though this is objectionable, as it may cause the brasses to warp or crack. Putting water on a very hot bearing should be resorted to only in an emergency, that is, when an engine *must* be kept running.

Water may be used on a moderately hot bearing without doing very much harm. It is quite common in practice, when sprinklers are fitted to an engine, to run a light spray of water on the crankpins when they show a tendency to heat, with very beneficial results.

Dangerous Heating.—Should a bearing become so hot as to scorch the hand or to burn oil before it is discovered or because of the necessity of keeping the engine running from some cause, it is imperative that the engine should be stopped, at least long enough to loosen up the brasses, even though it is necessary to start up again immediately; otherwise the brasses will be damaged beyond repair and deep grooves will be cut into the journals. If the brasses are babbitted, the white metal will melt out of the bearing at this stage. The engine will then be disabled, and if there is not a spare set of brasses on hand, it will be inoperative until the old brasses are rebabbitted or until a new set is made and fitted.

Running Engine **With** Hot Bearing.—If it is absolutely necessary in an emergency to keep the engine running while a bearing is very hot, the engineer must exercise his best judgment as to how he shall proceed. After slacking off the brasses, about the best he can do is deluge the inside of the bearing with a mixture of oil and graphite, sulphur, soapstone, etc., and the outside with cold water from buckets, sprinklers, or hose, taking the chances of ruining the brasses and cutting the journal.

Refitting Cut Bearing.—The wearing surfaces of the brasses and journal must be smoothed off as well as circumstances will permit; but if the grooves are very deeply cut, it will be useless to attempt to work them out entirely, and if the brasses are very much warped or badly cracked, it will be best to put in spare ones, if any are on hand. If not, the old ones must be refitted and used until a new set can be procured. As for the journal, it is permanently damaged. Temporary repairs can be made by smoothing down the journal and brasses; but at the first opportunity the journal should be turned in a lathe and the brasses properly refitted or replaced with new ones.

Newly Fitted Bearings.—The bearings of new engines are particularly liable to heat, as the wearing surfaces of the brasses and journal have just been machined and hence are comparatively rough. The conditions just mentioned also exist with new brasses and the journals of an old engine. If a new engine or one with new brasses is run moderately, in regard to both speed and load, and with rather loose brasses, there will be little danger of hot bearings, provided proper attention is given to adjustment and lubrication. This is what is familiarly termed *wearing down the bearings.*

Brasses Too Tight.—When the brasses of an engine bearing are set up too tight, heating is inevitable. It is often the case that an attempt is made to stop a pound in an engine by setting up the brasses when the thump should be stopped in some other way. The brasses should be slacked off as soon as possible. As a matter of fact, hot bearings should never occur from this cause.

Brasses Too Loose.—Bearings may heat because the brasses are too loose. The heating is caused by the hammering of the journal against the brasses when the crankpin is passing the dead centers. The derangement is easily remedied, however, by setting up the cap nuts or the key. Most engineers have their own views regarding the setting up of bearings. One method is to set up the cap nuts or key nearly solid and then slack them back half way; if the brasses are still too loose, they are set up again and slacked back less than before, repeating this operation until there is neither thumping nor heating.

Another method of setting up journal brasses is to fill up the spaces between the brasses with thin metal liners, from 18 to 22 Birmingham wire gauge in thickness, and a few paper liners for fine adjustment. Enough of these should be put in to cause the brasses to set rather loosely on the journal when the cap nuts or keys are set up solid. The engine should be run for a while in that condition; then a pair of the liners should be removed and the brasses set up solid again. This operation should be repeated until there is neither thumping nor heating. It may require a week or more, and with a large engine longer, to reach the desired point. If this system is carefully carried out, there will be very little danger of heating.

In removing the liners, great care should be exercised not to disturb the brasses any more than is absolutely necessary.

Warped and Cracked Brasses.—Warped and cracked brasses will cause heating, because they do not bear evenly on the journal, and hence the friction is not distributed evenly over the entire surface. If the distortion is not too great, the brasses may be refitted to the journal by chipping, filing, and scraping; but if they are twisted so much that they cannot, within reasonable limits, be refitted, nothing will do but new brasses.

Cut Brasses and Journals.—Brasses and journals that have been hot enough to be cut and grooved are liable to heat up again any time on account of the roughness of the wearing surfaces. As long as the grooves in the journal are parallel and match the grooves in the brasses, the friction is not greatly increased; but if a smooth journal is placed between brasses that are grooved and pressure is applied, the journal crushes the grooves in the brasses and becomes brazed or coated with brass, and then heating results. The way to prevent heating from this cause is to work the grooves out of the journal and brasses by filing and scraping as soon as possible after they occur.

Imperfectly Fitted Brasses.—Faulty workmanship is a common cause of the heating of crankpins, wristpins, and bearings. The brasses in that case do not bear fairly and squarely, even though they appear all right to the eye. A crankpin brass must fit squarely on the end of the connecting-rod and the rod itself must be square. If the key, when driven, forces the brasses to one side or the other and twists the strap on the rod, it will draw the brasses slantwise on the pin and make them bear harder on one side than on the other, thus reducing the area of the bearing surfaces. The same is true of the shaft bearings. If the brasses do not bed fairly on the bottom of the pillow-block casting or do not go down evenly, without springing in any way, heating will result.

Edges of Brasses Pinching Journal.—Brasses, when first heated by abnormal friction, tend to expand along the surface in contact with the journal; this would open the brass and make the bore of larger diameter were it not prevented by the cooler part near the outside and by the bedplate itself. If the brass has become hot quickly and excessively, the resistance to expansion produces a permanent set on the layers of metal near the journal, so that on cooling, the brass closes and grips the journal. This is why some bearings always run a trifle warm and will not work cool. A continuance of heating and cooling will set up a bending action at the middle of the brass, which must eventually end in cracking it. Heating produced in this way may be prevented by chipping off the brasses at their edges parallel to the journal, as shown at *a* in the aecompanying illustration, in which *b* is ·a section of the journal and *c* and *d* represent the top and bottom brasses, respectively.

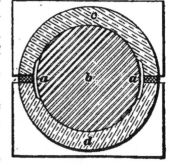

Stopped Oil Feed.—It does not take long for a bearing to get very hot if it is deprived of oil. The two principal causes of dry bearings are an oil cup that has stopped feeding, either by reason of being empty or by being clogged up from dirt in the oil, and oil holes and oil grooves stopped up with dirt and gum.

Insufficient Oil.—The effect produced upon a bearing by an insufficient oil supply is similar to that of no oil, but in a less degree. Of course, it will take longer for a bearing to heat with insufficient oil than with none at all, and the engineer has more time in which to discover and remedy the difficulty.

Dirty and Gritty Oils.—Oils that contain dirt and grit are prolific sources of hot bearings. There is a great deal of dirt in lubricating oils of the average quality; therefore, all oil should be strained through a cloth or filtered, no matter how clear it looks. All oil cups, oil cans, and oil tubes and channels should be cleaned out frequently. Oil may be removed from the cups by means of an oil syringe, and all oil removed from the cups and cans should be 'strained or filtered before being used.

Oils of Poor Quality.—There are on the market many lubricating oils whose quality cannot be definitely decided on without an actual trial, and it is a difficult matter to avoid getting a bad lot of oil sometimes. About the only safe way to meet this trouble is to pay a fair price to a reputable dealer for oil that is known to be of good quality, unless the purchaser is expert in judging oils.

Oil Squeezed Out of Bearings.—Bearings carrying very heavy shafts sometimes refuse to take the oil; or, if they do, it is squeezed out at the ends of the brasses or through the oil holes, and then the journal will run dry and heat. Large journals require oil of a high degree of viscosity, or heavy oil, as it is popularly called. Oil of this character has more difficulty in working its way under a heavy shaft than a thin oil has, but thin oil has not the body necessary to lubricate a large journal.

This difficulty may be met by chipping oil grooves or channels in the brasses. A round-nosed cape chisel, slightly curved, is generally used for this purpose; care should be taken to smooth off the burrs made by the chisel, which may be done with a steel scraper or the point of a flat file. The grooves are usually cut into the brass in the form of a **V** if the engine is required to run in only one direction; if it is to run in both directions the grooves should form an **X**. In the first instance, care must be taken that the **V** opens in the direction of rotation of the shaft; that is, the grooves should spread out from their junction in the same direction as that in which the journal turns. The oil grooves may be about ¼ in. wide and ⅛ in. deep, and semicircular in cross-section.

Grit in Bearings.—Grit is an ever-present source of heating of bearings, and only by persistent effort can the engineer keep machinery running cool in a dirty atmosphere. The machinery of coal breakers, stone crushers, and kindred industries is especially liable to be affected in this way. Work done on a floor over an engine shakes dirt down upon it at some time or other; hence, all floors over engines should be made dust-proof by laying paper between the planks. If the engine room and firerooms communicate, and piles of red-hot clinkers and ashes are deluged with buckets of water, the water is instantly converted into a large volume of steam, carrying with it small particles of ashes and grit that penetrate into every nook and cranny, and these will find their way into the bearings sooner or later. Hot clinkers and ashes should be sprinkled, and the fireroom door should be closed while the ashes and clinkers are being hauled or wet down or while the fires are being cleaned or hauled. As an additional precaution, all open oil holes should be plugged with wooden plugs or bits of clean cotton waste as soon as possible after the engine is stopped, and should be kept closed until ready to oil the engine again preparatory to starting up. Plaited hemp or cotton gaskets should also be laid over the crevices between the ends of the brasses and the collars of the journals of every bearing on the engine and kept there while the engine is standing still.

Overloading of Engine.—The effect produced by overloading an engine is this: the pressure on the brasses is increased to a point beyond that for which they were designed, the friction exceeds the practical limit, and the bearing heats. In case an engine is run at or near its limit of endurance, or if the journals are too small, it would be a wise and economical precaution to have a complete set of spare brasses on hand ready to slip in when the necessity arises.

Engine Out of Line.—If an engine is not in line, the brasses do not bear fairly upon the journals. This will reduce the area of the bearing surfaces in contact to such an extent as to cause heating. If the engine is not very much out of line, matters may be considerably improved by refitting the brasses by filing and scraping down the parts of those which bear most heavily on the journal. If this does not answer, the heating will continue until the engine is lined up.

The crosshead guides of an engine out of line are apt to heat, and they will continue to give trouble until the defect is remedied. The guides may also heat from other causes; for instance, the gibs may be set up too much. The danger of hot guides may be very much lessened by chipping zigzag oil grooves in their wearing surfaces and by attaching to the crosshead oil wipers made of cotton lamp wicking arranged so as to dip into oil reservoirs at each end of the guides if they are horizontal, and at the lower end if they are vertical. These wipers will spread a film of oil over the guides at every stroke of the crosshead.

16

Effect of External Heat on Bearings.—Bearings may get hot by the application of external heat. This may be the case if the engine is placed too near furnaces or an uncovered boiler, or in an atmosphere heated by uncovered steam pipes or other means. The excessive heat of the atmosphere will then expand the brasses until they nip the journals, which will generate additional heat and cause further expansion of the brasses, and so on until a hot bearing is the result. The remedy obviously depends upon the conditions of each case.

Brasses Too Long.—If the brasses are too long and bear against the collars of the journal when cold, they will most surely heat after the engine has been running a while. It is hardly possible to run bearings stone cold. They will warm up a little and the brasses will be expanded thereby, which will cause them to bear still harder against the collars. This, in turn, will induce greater friction and more expansion of the brasses. The evil may be obviated by chipping or filing a little off each end of the brasses until they cease to bear against the collars while running. A little side play is a good thing for another reason, which is that it promotes a better distribution of the oil and prevents the journal and brasses from wearing into concentric parallel grooves.

Springing of Bedplate.—If the bedplate of an engine is not rigid enough to resist the vibration of the moving parts, or if it is sprung by uneven settling or the instability of the foundation, the engine will be thrown out of line either intermittently or permanently, and the bearings will heat. But it will do no good to refit the brasses unless the engine bed is stiffened in some way and leveled up.

Springing or Shifting of Pillow-Block.—The effect of the springing or shifting of the pedestal or pillow-block is similar to the springing of the engine bed; that is, the bearing will be thrown out of line, with the consequent danger of heating. As the pedestal is usually adjustable, it is an easy matter to readjust it, after which the holding-down bolts should be screwed down hard. If a pedestal is not stiff enough to resist the strains upon it and it springs, measures should be taken to stiffen it.

STEAM TURBINES

ECONOMICAL CONSIDERATIONS

Steam Consumption.— The relation between the brake horsepower of the steam turbine at full load and the steam consumption is shown in the accompanying table. The values in this table are taken from published tests of steam turbines

STEAM CONSUMPTION PER HOUR OF TURBINES

Brake Horsepower	Pounds of Steam Used	Brake Horsepower	Pounds of Steam Used
100	18.2	600	15.3
200	17.5	700	14.8
300	16.9	800	14.3
400	16.3	900	13.7
500	15.8	1,000	13.2

that have attained the greatest commercial success. The turbines used saturated steam of from 115 to 140 lb. per sq. in., gauge pressure, and exhausted into a vacuum of from 26 to 28.5 in. of mercury. Better results than those noted in the table can be obtained by the use of highly superheated steam.

Effect of Vacuum.—The better the vacuum, the greater is the economy in the use of steam, both in the steam engine and in the steam turbine. A high vacuum is of greater value to the turbine, however, because the turbine can take advantage of a greater range of expansion. The degree of vacuum to be carried is a matter of dollars and cents; that is, it may cost more to create and maintain a high vacuum than may be saved in steam consumption. In a comparative test of a turbine and a triple-expansion engine under like conditions, it was found that, in the case of the reciprocating engine, little or nothing was to be gained by carrying a greater vacuum than about 26 in.; but the economy of the turbine in the use of steam increased rapidly as the vacuum was increased above 26 in.   The conclusion is that high degrees of vacuum are more desirable for turbines than for engines.

Advantages of Turbines.—The steam turbine possesses the following advantages over the reciprocating engine:

1.   The ability to use highly superheated steam, resulting in greater economy.

2.   Reduced cost per unit capacity of the electric generator because of increased speed and smaller weight per horsepower.

3.   Reduced floor space, resulting in less cost for land and power-station building.

4.   Reduced cost of lubrication, as no cylinder oil is needed and less oil is used for bearings.

5.   Saving in labor, as no oilers are required and one engineer can attend to more output than on reciprocating engines.

6.   Reduced cost of foundations, as the turbine is balanced and has no reciprocating parts.

7.   Good steam economy over a wider range of load than the reciprocating engine.   This is particularly advantageous in power stations, where the load is variable; thus, a turbine can be operated at one-fourth or one-half load with smaller increase of steam consumption than would be the case with a reciprocating engine under the same conditions.   Also, the turbine is more efficient than the engine under overloads.

The foregoing advantages apply mainly to turbines used on land.   The steam turbine, however, is also used for the propulsion of ships.   Among the advantages claimed for it in this class of service are the following:

1.   For the same power, the turbine plant has less weight than the engine installation.

2.   There is less danger of breakdowns, because of the smaller number of parts in a turbine installation.

3.   The balancing obtained almost eliminates vibration.

4.   With fast vessels, it is possible to obtain a higher speed than with reciprocating engines, at the same steam consumption.

5.   Less headroom is required than with steam engines.

Comparison of Turbines **and** Engines.--If the matter of steam consumption alone is considered, the average condensing turbine of less than about 700 H. P. is not so economical as the average compound or triple-expansion condensing engine, although the turbine may be preferred to the engine for other reasons. In larger sizes, however, and particularly in very large units, the economy of the turbine is very noticeable. The turbine possesses the ability to expand the steam to the lowest available condenser pressure without difficulty; but to do this in a reciprocating engine would require very large valves, and ports and heavy pistons, because of the great volume of steam to be handled at very low pressures.

Finding Horsepower of Turbines.—There is no way of finding the indicated horsepower of a steam turbine, because no form of indicator applicable to the turbine has been invented. Nor is any such instrument likely to be developed, owing to the very great difficulty of determining the energy given up to the blades of a turbine from a jet of steam. The usual way of finding the power of a steam turbine is to use a brake or a dynamometer and thus to determine the brake horse-power, or else to connect an electric generator to the turbine and measure the electrical output at the switchboard. In case the latter method is used, the efficiency of the generator and the turbine together is involved.

TURBINE TROUBLES

Clearance of Blades.—To obtain free running, it is necessary to allow clearance between the stationary and the moving rows of blades, as well as between the ends of the blades and the casing or the rotor. In impulse turbines, such as the Curtis and the Rateau, the clearance between the rows of blades is important; however, if it is made no greater than is necessary for mechanical reasons, the efficiency will not be affected seriously. In the reaction turbine, such as the Parsons, the clearance between the rows is of small consequence as compared with the clearance between the ends of the stationary blades and the rotor and between the ends of the moving blades and the casing. The former may vary from $\frac{1}{4}$ to 1 in. or more from the high-pressure to the low-pressure stage; but the tip clearance must be kept between a few hundredths and a few thousandths of an inch.

Stripping of Blades.—The stripping of the blades is one of the troubles to which turbines are subject. It may be due to the interference of the stationary and the movable blades, or to the rubbing of the blades against the shell or the rotor. In either of these cases the existing clearances are reduced, by wear of the parts, shifting of the rotor, or unequal expansion of the rotor and the casing, until the blades touch and tear one another loose. The same result will occur if some foreign solid, as a stray nut or bolt, is carried along with the steam into the turbine. If a turbine is started too quickly, without being properly warmed up, the sudden unequal expansion set up in the heavy casing and the lighter rotor may cause the blades to come in contact and be stripped. Stripping is claimed by some engineers to be more common in turbines in which the blades are not supported at their outer ends. To prevent it, therefore, shroud rings and metal lacings are applied to the blades at their outer ends, by some manufacturers of steam turbines.

Erosion of Blades.—As there are no valves, pistons, or piston rings in the turbine to be maintained free from leakage, about the only thing that can affect the steam consumption is the condition of the blades. The blades of steam turbines are subjected to the cutting action of steam flowing at high velocities, and often carrying water particles with it. This cutting, or erosion, wears away the edges and surfaces of the blades. From the data available, it appears that the erosion is very slight if the steam is dry or superheated, no matter what velocities are used; but if the steam is wet, erosion will take place, and it will be greatly increased if the velocity of the steam is high. The horsepower is not affected to any great extent by blade erosion, according to the results of experience. In the case of a 100-H. P. De Laval turbine, the steam inlet edges of the blades were worn away about $\frac{1}{16}$ in., yet the steam consumption was only about 5% above that with new blades.

Slugs of Water.—If the boiler supplying steam to a reciprocating engine primes badly, a slug of water may be carried
over into the cylinder, resulting in a cracked piston or cylinder,
a buckled piston rod or connecting-rod, or a wrecked frame.
In case a steam turbine is used, however, the danger is greatly
lessened. In turbines in which the blades are not supported
at their outer ends, the water may cause stripping of the blades;
but this is not very likely, as the blades at the high-pressure
end of the turbine are short. A rush of water from the boiler
has been known to bring a turbine almost to a stop without
damaging the blades.

Vibration.—On account of the high speeds attained in turbine practice, the rotors are balanced accurately, so as to reduce
vibration. But in spite of this careful balancing, vibration
may manifest itself during ordinary running. It may be caused
in any one of several ways, but the fundamental cause is lack
of balance. If the rotor is warmed up too rapidly, the shaft
or the wheels may be warped by unequal expansion, producing
an unbalanced effect. The stripping of a blade or two will
affect the balance of the wheel and tend to produce vibration.
Even water carried into the turbine with the steam will bring
about an unbalanced condition and will lead to vibration.
When vibration is observed, it is well to reduce the speed a
little, and to note whether this causes the vibration to cease.
If it does, but comes back again as soon as the speed is increased,
the source of the trouble should at once be determined.

OPERATION OF TURBINES

Inspection.—If the steam turbine is a new one, or if it has
been standing idle for a long period, it should not be started
until it, together with its auxiliary apparatus, has been thoroughly inspected. The bearings should be properly adjusted
and free from dirt, and the entire lubricating system should
be clean and filled with clean oil. The steam pipe from the
boilers should be blown through, so as to clear it of any foreign
matter that could be carried into the turbine by the steam.
The governor mechanism should be examined, to see that it

is in good order; the oil pump should be looked after, to ascertain whether it is in condition to maintain a continuous supply of oil; and, finally, before the turbine is started, the shaft should be turned over by hand, to insure that the rotor will turn freely in the casing.

Starting.—A steam turbine should be started slowly, and before it is allowed to turn over under steam it should be warmed up. This is accomplished by opening the throttle valve just enough to let steam flow into the turbine. The drains should be kept open until the turbine is well started. The length of time required for warming up depends on the size of the turbine, a large unit requiring more time than a small one. As the warming up proceeds, the throttle may gradually be opened more, and the auxiliary machinery may be started. Once it has been started, the turbine should be brought up to speed slowly. If it is speeded up too rapidly, vibration will result. After the normal running speed has been reached, the load may be thrown on; but this, also, should be done gradually, to prevent a rush of water from the boiler with the steam.

If superheated steam is used, extra caution must be employed in starting, for during the warming up, with the throttle valve only slightly opened, the passing steam will be cooled considerably. But when the valve is opened wider, the greater volume passing will not lose so much of its superheat, and if care is not exercised the turbine will be subjected to sudden expansion because of the higher temperature of the steam. The main point in starting is to avoid any sudden changes of temperature in the turbine. If a turbine must be ready to be put in operation at short notice, steam may be allowed to flow through it continually, by means of a by-pass around the throttle valve. It will always be warmed up, then, and can be brought up to speed with less danger and more rapidly.

Lubrication of Bearings.—The shaft or spindle of a turbine rotates at high speed, and therefore the bearings should be kept well lubricated; for if the oil supply fails, or if a bearing begins to heat because of grit carried into it, the resulting trouble will come very quickly. The presence of a hot bearing will usually be evidenced by the smell of burning oil or by the appearance of white smoke. When these signs are observed the oil supply should immediately be increased to the greatest possible amount. If this does not reduce the temperature of the bearing or prevent its further heating, the turbine should be shut down. To continue will result in burning out the bearing, and it is better to stop before this happens. The high speed of the shaft renders it impossible to nurse a hot turbine bearing along as is done frequently in the running of reciprocating engines.

Use of Superheated Steam.—As there are no internal rubbing surfaces in the steam turbine, superheated steam may be employed without causing any of the lubrication troubles attending its use in reciprocating engines. Because of the greater amount of heat contained in a pound of superheated steam, the economy of a turbine working with superheated steam is greater than that of one working with saturated steam; also, the efficiency is increased because the superheated steam causes less frictional resistance to the motion of the blades. To show the value of superheated steam in turbine work, it may be stated that 50° F. of superheat reduces the steam consumption about 6%; 100° F. of superheat reduces it about 10%; and 150° F. of superheat reduces it about $13\frac{1}{2}\%$. The use of high superheat, however, produces expansion of the rotor and the casing and may cause the blades to interfere; as a result, the usual degree of superheat in steam-turbine practice is 100° F., and seldom exceeds 150° F.

Exhaust-Steam Turbine.—The steam turbine shows better economy than the steam engine when working with low-pressure steam in connection with a high vacuum; but when working with high-pressure steam and a vacuum of about 26 in., the engine is the more economical. As a consequence, a combination of the steam engine and the steam turbine has been adopted. The engine uses the high-pressure steam from the boilers and expands it to about atmospheric pressure. This exhaust steam then passes into the turbine, which exhausts into a condenser carrying a high degree of vacuum, and the expansion is carried to the extreme practicable limit. The turbine thus used in connection with an engine is termed an *exhaust-steam turbine.*

Maintenance of Vacuum.—As the economy of the steam turbine is dependent so largely on the degree of vacuum carried, it is necessary for the engineer to watch the vacuum gauge closely. With reciprocating engines, the loss of 1 or 2 in. of vacuum may not be of much consequence; but in a turbine plant, where the vacuum is from 27 to 28 in., a loss of 1 or 2 in. will result in a considerable increase in the steam consumption. Because of the high vacuum employed, the difficulty of keeping pipes, valves, and glands from leaking is greater in turbine practice than in engine practice, but the greater economy obtained by keeping everything tight overbalances the increased care and labor.

Shutting Down.—In shutting down a steam turbine, the throttle valve should be closed partly before the load is reduced, so as to prevent any possibility of racing when the load is finally taken off. The load may then be used as a brake to bring the rotor to a stop. When the throttle valve has been closed and the steam supply has been shut off completely, the auxiliary machinery may be stopped. If the load is taken off before the throttle is wholly closed, the turbine may continue to rotate for half an hour, as the rotor is then running in a vacuum and under no load. The speed may be reduced by opening the drains and allowing air to enter the casing. The oil supply to the bearings must be continued until the turbine has come to rest, and the oil pump should be the last auxiliary to be stopped.

Care of Gears in De Laval Turbines.—The De Laval Steam Turbine Company in their directions for operating their turbines state that in order to keep the gears in good condition the teeth should be cleaned occasionally when the turbine is not in service. They recommend that a wire brush and kerosene be employed for this purpose. At the same time the gear-case should also be thoroughly cleaned, and after the cleaning the gears should be well lubricated.

Should an engineer for any reason desire to take the gears out of the case, it is recommended that he secure special directions relating to their removal from the manufacturers. The same statement also applies to the adjustment of the gears, which need to be kept in perfect adjustment.

PROPULSION OF VESSELS

SLIP

True Slip.—In considering the speed of a stream projected by a propelling instrument from a ship or other vessel in motion, it must be borne in mind that while the stream is propelled astern the vessel is advancing. Since the stream must move astern faster than the vessel advances, the rearward speed of the stream in relation to a fixed point of the water some distance astern of the ship, as a floating piece of wood, will be the difference between the speed of the vessel in relation to the piece of wood and the rearward speed of the stream in relation to the vessel. When the propeller works in a wake, which has a forward motion, the speed with which water is fed to the propelling instrument is reduced thereby and it becomes equal to the difference between the forward speed of the vessel in relation to a fixed point of the water clear of the wake and the wake velocity. Thus, if the speed of the ship is 15 mi. per hr. and a wake that has a forward velocity of 3 mi. per hr. collects at the stern, the speed with which the water is fed to a screw propeller is $15-3=12$ mi. per hr. The difference between the speed with which water is fed to the propelling instrument and the speed with which it is projected astern, both speeds being measured in relation to the vessel, is called the *true slip*, and also the *real slip*, of the stream.

Apparent Slip.—In practice, it is very inconvenient and exceedingly difficult to measure either the wake velocity or the speed of the ship in relation to the wake, but it is a very simple matter to measure the speed of the vessel in relation to a fixed point in the water clear of the wake by means of an instrument called a *log*. The difference between the speed of the stream projected by the propelling instrument and the speed of the ship thus found is taken as the slip. When calculated in this manner it obviously is not the same as the true slip; it is called the *apparent slip*.

Formulas for Slip.—It is customary to express slip in per cent. of the velocity of the stream projected by the propelling instrument.

Let S_t = true slip in per cent., expressed decimally;

S_a = apparent slip in reference to the ship's motion through the water, expressed decimally and in per cent.;

V = velocity of stream projected by propelling instrument, in relation to vessel;

V_1 = velocity of water fed to propelling instrument; that is, speed of vessel in relation to the surrounding water diminished by the wake velocity at the point where the propelling instrument is located, for a vessel under way.

V_2 = speed of vessel in relation to the water, as shown by the log.

Then, the true slip is given by the formula

$$S_t = \frac{V - V_1}{V} \qquad (1)$$

and the apparent slip by the formula

$$S_u = \frac{V - V_2}{V} \qquad (2)$$

It is customary among writers on marine propulsion to refer to apparent slip simply as slip; when the true slip is meant, it is usually called distinctly the true slip.

SCREW PROPELLERS

Definitions.—If a point is caused to rotate at a uniform distance from and about an axis, and if the point at the same time is caused to advance at a uniform rate in the direction of axis, its path will be a *helix*. If the point, when moving away from the observer, moves in the direction of the hands of a watch, the helix will be *right-handed;* if in an opposite direction, *left-handed.* The distance that the point advances in one complete revolution is known as the *pitch.* If a line passing through the axis is caused to rotate about the axis, and to pass along the path of the point mentioned above, its path will be the surface of a *true screw*, provided the angle that the line makes with the axis remains constant. From this, it follows that a true screw is one in which the advance of any point, in the direction of the axis, at any distance from it, for any part of a revolution, but the same in each case, is the same. By causing lines making equal angles with each other and the axis to rotate about the axis in a helical path, a *multiple-threaded true screw* will be generated, having the same pitch as a single-threaded true screw generated by a line following the same helical path.

Consider a four-threaded, right-handed screw, generated by the lines *O A*, *O B*, *O C*, and *O D*, in the accompanying illustration. These lines represent the intersections of the four helical surfaces with a 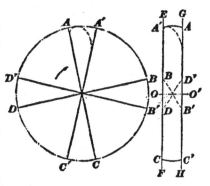 plane *E F* perpendicular to the axis. Assume the helical surfaces to be cut by a plane, as *G H*, parallel to the first and intersecting the axis at another point. Then, *O A A' O'*, *O B B' O'*, *O C C' O'*, and *O D D' O'* will be the helical surfaces of the blades of a four-bladed, right-handed screw propeller. If pieces of metal are shaped to conform to these helical surfaces and if these pieces of metal, which are called *blades*, are fastened to a hub, which in turn is keyed to a shaft rotated by an engine, the *screw propeller* in its simplest form is obtained. If the screw propeller is revolving in the direction of the arrow, the portion of the blade that strikes the water first, which will be near the plane *G H*, is known as the *anterior portion* of the blade; and the portion that is near the plane *E F*, as the *posterior portion*. That part of the blade that is near the periphery *A A'* is known as the *tip*. In practice, screw propellers are hardly ever made of the shape shown. Generally the anterior portion of the blade is rounded off toward the tip, as shown by the dotted line on the blade *O A `A' O'*. The posterior portion is also slightly rounded. Very often part of the anterior portion near the hub is also cut away.

Radially Expanded Pitch.—Sometimes the surfaces of the blades are not truly helical; as usually found, the pitch near the tip is greater than the pitch near the hub. Such a propeller is said to have a *radially expanded pitch*. The reason for constructing the blade in this manner is this: Since the part of the blade near the hub strikes the water at nearly a right angle, it acts chiefly to churn the water, and since the water near the periphery is thereby disturbed, the tip of the blade acts on water in motion. By increasing the pitch at the tip, it is supposed that the resistance at all parts of the blade is more nearly equalized.

Axially Expanded Pitch.—The blades are sometimes constructed in such a manner that the anterior portion of the blade has a finer pitch than the posterior portion. Such a blade is said to have an *expanding* or *axially expanded pitch*. The object to be attained by it is as follows: The anterior portion of the blade, striking on water at rest and encountering the resistance due to a solid body moving through water at rest, sets the water in motion, driving it astern. Therefore, the posterior portion acts on water in motion. By expanding the pitch to the same extent, further motion is given to the water by the posterior portion, and it is supposed that the resistance at all parts of the blade is thereby equalized, the same as with radially expanded pitch blades.

Surface Areas.—The actual area of the surface on the driving side of a propeller blade is known by various names, as the *developed blade area*, the *helicoidal blade area*, or simply the *blade area*. When referring to the total blade area, it is usually spoken of as the *developed propeller area*, the *helicoidal propeller area*, or simply, the *propeller area*. The area of a blade projected on a plane at right angles to the propeller shaft is called its *projected area;* the projected area of all the blades is the *projected propeller area*. The area of the circle described by the tips of the blades is the *disk area* of the propeller. The *pitch ratio* is the ratio of the pitch of the propeller to its diameter.

Measurement of Pitch.—In practice, the pitch of a propeller may be found quite closely in the manner shown in the illustration. Take a piece of joist or lath D, which should be as straight as possible and place it so as to touch one of the blades at any distance, as *b*, from the axis *A B*, taking care to hold it parallel to the axis. Next take a carpenter's square, shown at E, and place it on the lath and against the blade, so that the point at which the square touches the blade will be the same distance from the axis as is the lath. Measure the distances *a*, *b*, and *c*; *a* is the distance from the square to the point at which the lath touches the blade, and *c* the distance from the point at which the square touches the blade to the lath. The distances *a* and *c* may be obtained in a different

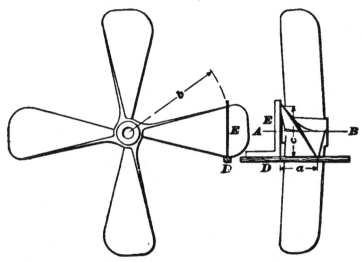

manner, if considered more convenient, thus: Place the screw propeller so that one blade is horizontal. To a piece of string about 10 ft., or more, in length tie two nuts; place the string over the blade, with the nuts hanging down, at the distance from the shaft axis at which it is desired to find the pitch, taking care to place the string so that both parts hanging down arc the same distance from the axis. The distance the two parts are apart is the distance *a*. To find *c*, hold a lath against the blade and both vertical parts of the string; while holding the lath parallel to the shaft axis the distance *c* can be measured.

The pitch of the propeller may then be calculated by the

formula $$P = \frac{6.2832\,ba}{},$$

in which P = pitch of screw propeller;

      a = depth of blade;

      b = distance from center of shaft where width and depth of blade is measured;

      c = width of blade.

The distances a, b, and c should all be taken in inches, and the measurements for pitch should always be taken on the side of the blade that strikes the water when propelling the vessel ahead.

EXAMPLE.—If a screw propeller blade 6 ft. from the center of the shaft is 22 in. deep and 41 in. in width, at right angles to the shaft, what is the pitch?

SOLUTION.—Applying the formula,

$$P = \frac{6.2832 \times 12 \times 6 \times 22}{41} = 242.75 \text{ in.} = 20 \text{ ft. } 2\tfrac{3}{4} \text{ in.}$$

Determining **the** Kind of Pitch.—To determine whether a screw propeller is a true screw, two or more measurements of the pitch should be taken on different parts of the blade at the same distance from the axis. Another set of measurements should be taken at some other distance from the axis. If the pitches calculated from these measurements agree closely, the propeller is a true screw.

To determine whether the pitch of the screw is radially expanded, calculate the pitch at two or more distances from the axis; if the pitch increases toward the tip of the blade, the screw propeller is of radially expanded pitch.

To determine whether the screw has an expanding pitch, the pitch must be calculated for the anterior and posterior portions of the blade. The pitch for the posterior portion should be the coarser; and, if calculated for any distance from the axis, the pitches of the anterior portion, as well as those of the posterior portion of the blade, should agree, provided that the axial measurements are taken in the same planes passing through the axis.

Required Pitch of Propeller.—The pitch required for a screw propeller may be found by the formula

$$P = \frac{V_2}{(1 - S_a)N},$$

in which P = pitch, in feet;

V_2 = speed of ship, in feet per minute;

S_a = apparent slip, in per cent., expressed decimally;

N = revolutions per minute;

Required Diameter of Propeller.—For the diameter of a screw propeller, Seaton gives the formula:

$$D = C\sqrt{\frac{H}{(P\,N)^3}},$$

in which D = diameter of screw propeller, in feet;

H = indicated horsepower;

P = pitch of screw propeller, in feet;

N = revolutions per minute;

C = a constant ranging from 17,000 for slow freight steamers to 25,000 for fast-running light steamers, as torpedo boats and fast steam launches.

EXAMPLE.—Find the diameter of a screw propeller for a steam launch with an engine of 10 H. P., the screw having a pitch of 4 ft. and making 200 R. P. M.

SOLUTION.—Applying the formula,

$$D = 25,000 \times \sqrt{\frac{10}{(4 \times 200)^3}} = 3.5 \text{ ft.}$$

When the rule gives a diameter that is impossible for the conditions, either P or N, or both, must be varied. Making either or both of these values larger will give a smaller diameter of screw; conversely, making either or both of these values smaller gives a larger diameter of screw. The rule is intended for screw propellers with four blades; if three blades are to be used, the diameter should be increased about 10%; and if two blades are to be used, about 20%. The pitch ratio varies in practice between 1.1 and 1.6.

Blade Area.—The total blade area of four-bladed propellers ranges from 35 to 45% of the disk area; in three-bladed propellers, it ranges between 27 and 33%, and in two-bladed propellers, between 20 and 25%. The value to be chosen should vary with the pitch ratio, using a low total blade area for a low pitch ratio and increasing the value as the pitch ratio is made greater.

17

SPEED OF VESSELS

Powering of Vessels.—The exact power required to propel a vessel at a given speed cannot be found very readily by the principles of mechanics. Instead, empirical rules based on the actual performance of vessels are usually relied on. The conditions that influence the relation between power and speed are many, but only a few of the more important ones will be enumerated here. For instance, the area of the blades of the screw propeller may not be sufficient for high speed, owing to a churning of the water when the propeller is revolved beyond a certain speed; and, although the power expended in revolving the propeller faster may be considerable, the increase of the speed of the vessel may be very slight. A similar state of affairs may occur if the area of the buckets of a paddle wheel is too small. It may be amply sufficient for a low rate of speed, and still be entirely too small for a higher rate, thus showing, probably, a high efficiency of the propelling instrument at a low speed, and a very poor one at a higher rate. Again, the efficiency of the engine may vary greatly for different powers developed by the same engine. Therefore, no hard-and-fast rule that will express the relation between power and speed under all conditions can be laid down.

Admiralty Rule.—The rule most frequently used in the powering of vessels is known as the *Admiralty rule.* It involves the selection of a proper constant based on actual experience, when this constant, a number of which are given in the accompanying table, is properly selected, the results of the rule will be found to agree very closely with the actual performance of vessels powered by the rule, at least under ordinary conditions and for ordinary efficiencies of the propelling apparatus. The formula is

$$H = \frac{S^3 \cdot \sqrt[3]{W^2}}{k},$$

in which H = indicated horsepower;

W = displacement of vessel, in tons of 2,240 lb.;

k = a constant taken from the table;

S = speed, in knots.

Determining Fineness of Vessel.—To determine whether a vessel is fair or fine, its displacement, in cubic feet, is usually compared with the volume of a rectangular box having a length equal to the length of the vessel on the water-line, a width equal to the beam, and a depth equal to the mean draft of the vessel diminished by the depth of the keel. If the displacement is .55 of the volume of the box, or less, the vessel is fine; if above .55 and less than .70, fair. The quotient obtained by dividing the displacement by the contents of the imaginary box is called the *coefficient of fineness*.

VALUES OF *k* IN ADMIRALTY RULE

Description of Vessel	Speed Knots	*k*
Under 200 ft., fair...................	9 to 10	200
Under 200 ft., fine...................	9 to 10	230
Under 200 ft., fine...................	10 to 11	210
Under 200 ft., fine...................	11 to 12	200
From 200 to 250 ft., fair.............	9 to 11	220
From 200 to 250 ft., fine.............	9 to 11	240
From 200 to 250 ft., fine.............	11 to 12	220
From 250 to 300 ft., fair.............	9 to 11	250
From 250 to 300 ft., fair.............	11 to 13	220
From 250 to 300 ft., fine.............	9 to 11	260
From 250 to 300 ft., fine.............	11 to 13	240
From 250 to 300 ft., fine.............	13 to 15	200
From 300 to 400 ft., fair.............	9 to 11	260
From 300 to 400 ft., fair.............	11 to 13	240
From 300 to 400 ft., fine.............	11 to 13	260
From 300 to 400 ft., fine.............	13 to 15	240
From 300 to 400 ft., fine.............	15 to 17	190
Above 400 ft., fine...................	15 to 17	240

Selecting Constant in Admiralty Rule.—The selection of a proper value of *k* calls for the exercise of considerable judgment, based on personal knowledge of the actual performance of similar vessels. Generally speaking, the value of *k* is influenced by the length, speed, and shape of the vessel. The value of *k* should be greater with an increased length of the vessel in proportion to the width, and also with a finer under-water body; conversely, its value should be less as the ratio of length to width becomes smaller, and as the form becomes

fuller. Furthermore, the value of k should be smaller for
relatively high speeds than for low speeds, for vessels of the
same form and displacement.   Prof. W. F. Durand states that
a speed may be considered as relatively high or low if the
speed exceeds the numerical value of the square root of the
length of the vessel in feet, or falls below it.   Thus, if a vessel
is 64 ft. long, $\sqrt{64} = 8$, a speed of 10 knots would be considered
as relatively high, while a speed of 5 knots would be considered
as relatively low.   For small boats, if the speed is given in
statute miles per hour, the values of k range between 150 and
225, and for speeds given in knots, between 100 and 150,
according to Prof. W. F. Durand.

   Relation of Horsepower and Revolutions.—The speed of a
ship fully under way is about directly proportional to the
number of revolutions made by the engine.   But the power
required to turn the propelling instrument varies as the cube
of the number of revolutions, or, what is the same thing, as
the cube of the speed.   Hence, it is possible to find, approximately, the power developed by an engine for any given
number of revolutions per minute, any other horsepower and
the corresponding revolutions per minute being known by the
use of the formula

$$H_1 = \frac{R_1^3 H}{R^3},$$

in which H_1 = required indicated horsepower;
            R_1 = revolutions per minute at required power;
            H = given indicated horsepower;
            R = revolutions corresponding to the horsepower H.

   In practice, the horsepower calculated by this formula will
not always correspond to that actually used, as found by the
indicator diagram.   This is due to the fact that the efficiency
of the machinery is not necessarily the same at all speeds.
As a general rule, the engine will be at its maximum efficiency
at some certain speed, and will have a lower efficiency at a
higher or lower speed.   The speed at which the engine is at
its best efficiency can be found only by actual trial.

   Reduction of Horsepower When Towing.—It is a well-known
fact among marine engineers that an engine will develop a lower
horsepower with a given boiler pressure, throttle position,

and cut-off when towing or having the resistance of the vessel increased by other means, than will be developed under the same engine conditions but running free. The reason for this is explained in the following discussion, in which for the sake of simplicity two convenient assumptions have been made that are not absolutely correct in practice. These assumptions are that the horsepower of an engine varies directly as the number of revolutions, the mean effective pressure remaining the same, and that the speed of the vessel varies directly as the number of revolutions.

Consider a paddle-wheel steamer running free, with its engine developing its greatest horsepower possible. Since the turning effort of the engine depends on only the mean effective pressure in the cylinders, it is independent of the revolution so long as the throttle position, boiler pressure, and cut-off remain the same. This turning effort, when exerted at the circumference of the effective diameter circle tangentially to the same parallel to the surface of the water, is the total force tending to propel the vessel forwards, and is resisted by an opposing force, which·is the resistance of the vessel. Let the engine be started and assume that it is making its greatest turning effort. The resistance being less than the forward force, the vessel moves forwards under the influence of a forward accelerating force equal to the difference between the total forward force and the resistance. As the vessel gathers headway, the resistance increases; this means that the difference between the forward and the resisting force, that is, the accelerating force, decreases until the total forward force and total resistance have become equal, when the vessel continues at a uniform speed. Let the resistance be increased either by the vessel picking up a tow, by a head-wind or head-sea, by encountering an adverse current, or by a combination of these circumstances. The conditions remaining the same as before at the engine, the turning effort, that is, the total forward force,·is the same; but, as the initial resistance is increased, the initial difference between the total forward force and the total resistance is smaller than in the first case. This means that a smaller accelerating force is available with an increased resistance, and consequently the total forward force and total resistance become equal at a lower speed of the vessel, which then continues under way at a uniform, but lower speed. Now, the horsepower of an engine varies (theoretically) directly as the number of revolutions, the mean effective pressure remaining constant. It has been shown that the speed of the vessel has been lowered; from this it follows that the revolutions and consequently the horsepower must be less when the resistance has been increased. By adding to the resistance of the vessel, a condition is finally reached similar to that of a vessel moored to a dock; the forward force and resistance are equal and the vessel, as no accelerating force is available, remains stationary. In this condition, the number of revolutions, and, hence, the horsepower, has dropped to the lowest limit.

Relation of Coal Consumption to Speed.—The fuel consumption may be said to vary directly as the horsepower developed (this is not exactly true, but only approximately). The horsepower varies about directly as the cube of the speed, whence it follows that the fuel consumption will also vary as the cube of the speed, approximately. Hence, to find the probable coal consumption for a speed different from a known speed, use the formula

$$c = \frac{s^3 C}{S^3}, \qquad (1)$$

in which $c =$ coal consumption, in tons, at the new speed;
  $s =$ new speed;
  $S =$ known speed;
  $C =$ coal consumption, in tons, at the known speed.

To find approximately the speed of steaming for a new coal consumption, use may be made of the formula

$$s = \sqrt[3]{\frac{c\, S^3}{C}} \qquad (2)$$

At sea, owing to an accident, it is often desired to know what speed to maintain in order to reach a given port with the amount of coal on hand. This problem is readily solved by trial and by application of formula 1. In practice, a good margin of coal should be shown by the calculations as left over, for the reason that the actual coal consumption at the

reduced speed will, as a general rule, be in excess of the calculated consumption, by reason of the decrease in economy of
the engine induced by reducing the developed horsepower.

EXAMPLE.—A steamer consumes 20 T. of coal per day at
a normal speed of 10 knots; the distance to the nearest port
where coal can be had is 600 mi., and the estimated quantity
of coal in the bunkers is but 35 T. Find what speed should
be maintained in order to reach the coaling station with the
coal supply on hand.

SOLUTION.—The best way to proceed in a case of this kind
is to assume a lower speed, as 8 knots, and calculate the new coal
consumption for that speed; thus, $c = \dfrac{8^3 \times 20}{10^3} = 10.24$ T. per
da., or .43 T. per hr. The time required to cover a distance
of 600 mi. at a speed of 8 knots is $600 \div 8 = 75$ hr., and at a
coal consumption of .43 T. per hr. the total quantity of coal
required at that speed is $75 \times .43 = 32.25$ T. Hence, if a speed
of 8 knots is maintained, the supply of coal on hand, or 35 T.,
will suffice to reach the coaling station under ordinary weather
conditions.

Relation Between Engine Speed and Ship's Speed.—On
taking charge, the number of revolutions to produce a given
speed is often desired. In that case the pitch of the screw, or
the effective diameter of the paddle wheel (taking the effective
diameter for this purpose from center to center of buckets)
must be measured and a fair slip value assumed. Then, to
find the revolutions per minute, multiply the pitch of the
screw in feet, or the circumference in feet of the effective
diameter circle of a paddle wheel, by 60, and by the difference
between 1 and the assumed apparent slip expressed decimally.
Divide the speed of the ship in feet per hour by this product.

EXAMPLE.—The pitch of a screw propeller is 16 ft.; how
many revolutions per minute must it make to drive the ship
at the rate of 10 knots, the apparent slip being estimated
at 10%?

SOLUTION.—Applying the rule and taking the knot at
6,080 ft.
$$\frac{10 \times 6,080}{60 \times 16 \times (1 - .1)} = 70.37 \text{ rev. per min.}$$

ENGINEERS' LICENSE LAWS

STATES AND CITIES HAVING LICENSE LAWS

Engineers' license laws, on January 1, 1913, are in force in the following states and cities in the United States of America:

States.—Massachusetts, Minnesota, Montana, Ohio, Pennsylvania, and Tennessee.

Cities.—Alleghany, Pa.; Atlanta (Fulton County), Ga.; Baltimore, Md.; Buffalo, N. Y.; Chicago, Ill.; Denver, Colo.; Detroit, Mich.; Elgin, Ill.; Goshen, Ind.; Hoboken, N. J.; Huntington, W. Va.; Jersey City, N. J.; Kansas City, Mo.; Lincoln, Neb.; Los Angeles, Cal.; Memphis, Tenn.; Milwaukee, Wis.; Mobile, Ala.; New Haven, Conn.; New York, N. Y.; Niagara Falls, N. Y.; Omaha, Neb.; Peoria, Ill.; Philadelphia, Pa.; Pittsburg, Pa.; Rochester, N. Y.; Saginaw, Mich.; Santa Barbara, Cal.; St. Joseph, Mo.; St. Louis, Mo.; Scranton, Pa.; Sioux City, Ia.; Spokane, Wash.; Tacoma, Wash.; Terre Haute, Ind.; Washington (District of Columbia); and Yonkers, N. Y.

LICENSE LAWS FOR STATIONARY ENGINEERS

ABSTRACTS OF STATE LAWS

Following are abstracts of the license laws and ordinances of the different states and most of the cities just named. Abstracts of license ordinances of some cities are omitted because copies of the ordinances of these places were not available at the time of going to press.

Massachusetts.—In Massachusetts, the engineers' and firemen's license law is under the supervision of the District Police, Boiler-Inspection Department, State House, Boston.

Section 78, chapter 102, of the revised laws, acts of 1907, and amendments, which took effect January 1, 1912, reads:

"No person shall have charge of or operate a steam boiler or engine in this commonwealth, except boilers and engines upon locomotives, motor road-vehicles, boilers in private residences, boilers in apartment houses of less than five flats, boilers under the jurisdiction of the United States, boilers for agricultural purposes exclusively, boilers of less than 9 H. P., and boilers used for heating purposes exclusively, which are provided with a device approved by the chief of the district police limiting the pressure carried to 15 lb. per sq. in., unless he holds a license as hereinafter provided."

A steam boiler or engine may not be operated for more than 1 wk., unless the person in charge of and operating it is duly licensed.

"Section 80.—The words *have charge* or *in charge*, used in the foregoing section, shall designate the person under whose supervision a boiler or engine is operated. The person operating shall be understood to mean any and all persons who are actually engaged in generating steam in a power boiler." This section of the Massachusetts laws has been amended so as to include a designation of the terms *operator, operated,* or *operating*, where used in the law, as applying to any person who, under the supervision of the licensed person in charge, operates any appurtenances of a boiler or engine, provided that there is not more than one such person employed for every licensed person, and that any such operating must be in the presence of and under the personal supervision of the latter person.

Section 81 provides that whoever desires to act as engineer or fireman shall apply for a license therefor to the state inspector of boilers for the city or town in which he resides. The application blanks are to be obtained from the boiler-inspection department of the district police.

To be eligible for a first-class fireman's license, a person must have been employed as a steam engineer or fireman in charge of or operating boilers for not less than 1 yr., or he must have held and used a second-class fireman's license for not less than 6 mo. To be eligible for examination for a third-class engineer's license, a person must have been employed as a steam engineer or fireman in charge of or operating boilers for not less than 1½ yr., or he must have held a first-class fireman's license for not less than 1 yr. To be eligible for examination for a second-class engineer's license, a person must have been employed as a steam engineer in charge of a steam plant or plants having at least one engine of over 50 H. P., for not less than 2 yr.; or he must have held and used a third-class engineer's license for not less than 1 yr., or have held and used a special license to operate a first-class plant for not less than 2 yr.; except that any person who has served 3 yr. as apprentice to the machinist in charge or boilermaking trade in stationary, marine, or locomotive engine or boiler works, and who has been employed for 1 yr. in connection with the operation of a steam plant, or any person graduated as a mechanical engineer from a duly recognized school of technology, who has been employed for 1 yr. in connection with the operation of a steam plant, shall be eligible for examination for a second-class engineers' license.

To be eligible for examination for a first-class engineer's license, a person must have been employed for not less than 3 yr. as a steam engineer in charge of a steam plant or plants having at least one engine of over 150 H. P., or he must have held and used a second-class engineer's license, in a second-class or first-class plant, for not less than 3½ yr. The applicant shall make oath to the statements contained in his application.

Section 82 provides that licenses shall be granted according to the competence of the applicant and shall be distributed in the following classes:

Engineers' licenses: First class, to have charge of and operate any steam plant; second class, to have charge of and operate a boiler or boilers, and to have charge of and operate engines, no one of which shall exceed 150 H. P., or to operate a first-class plant under the engineer in direct charge of the plant; third class, to have charge of and operate a boiler or boilers not exceeding in the aggregate 150 H. P., and an engine not exceeding 50 H. P., or to operate a second-class plant under the engineer in direct charge of the plant; fourth class, to have charge of and operate hoisting and portable engines and boilers.

Under the heading of classification of licenses, two new classes have recently been added, namely, a portable class and a steam fire-engine class. The former applies to a person having charge of or operating portable boilers and engines, except hoisting engines and steam fire-engines, and the latter applies to a person who has charge of or operates a steam fire-engine or boiler.

Minnesota.—The engineers' license and boiler-inspection laws of the state of Minnesota are administered by a board of inspectors, one of whom shall reside in each senatorial district. This board is appointed biennially by the governor of the state. It also has jurisdiction over the vessels on the inland waters of the state, in relation to masters and pilots, as well as to engineers and the machinery of such vessels.

The laws state that engineers shall be divided into four classes, namely, chief engineers, first-class engineers, second-class engineers, and special engineers.

1. No license shall be granted to any person under 21 yr. of age, except to special engineers. No license shall be granted to any person to perform the duties of chief engineer who has not taken and subscribed an oath that he has had actual experience of at least 5 yr. in operating steam boilers and steam machinery, or whose knowledge, experience, and habits of life are not such as to justify the belief that he is competent to take charge of all classes of steam boilers and steam machinery.

2. No license shall be granted to any person to act as first-class engineer who has not had actual experience of at least 3 yr. in operating steam boilers and steam machinery, and whose experience and habits of life are not such as to warrant the belief that he is competent to take charge of all classes of steam boilers and steam machinery not exceeding 300 H. P.

3. No license shall be granted to any person to act as second-class engineer who has not had at least 1 yr. of actual experience in operating steam boilers and steam machinery, or whose experience and habits of life are not such as to warrant the belief that he is competent to take charge of all classes of steam boilers and steam machinery not to exceed 100 H. P.

4. No license shall be granted to any person to act as special engineer unless he is found on examination to be sufficiently acquainted with the duties of an engineer to warrant the belief that he can be safely entrusted with steam boilers and steam machinery not to exceed 30 H. P.

The fee for the examination of an applicant for an engineer's license shall be $1; for the biennial renewal of certificates of license, the fee shall be $1, which fee shall accompany the application. Applicants must subscribe oath as to their experience period.

Montana.—In Montana, the license laws are administered by an inspector of boilers, who is appointed by the governor of the state. The inspector has an assistant, and they must, as often as is convenient, publish in some suitable newspaper a notice stating on what days they will be in certain specified localities. This notice must also state that they will, at the time and place specified in such notice, receive applications and make examination for the purpose of granting engineers' certificates and examine boilers subject to inspection.

In Montana, engineers entrusted with the care and management of steam machinery are divided into three classes, namely, first-class engineers, second-class engineers, and third-class engineers. A candidate for a first-class license must have had at least 3 yr. of actual experience in the operation of steam boilers and steam machinery, or his knowledge and experience must be such as to justify the belief that he is competent to take charge of all classes of steam boilers and steam machinery. A candidate for a second-class license must have had at least 2 yr. of experience in the operation of steam boilers and steam machinery, and must on examination be found competent to take charge of all classes of steam boilers and steam machinery not exceeding 100 H. P. A candidate for a third-class license must have served at least 1 yr. as fireman under a competent engineer, and must be, on examination, found competent to be entrusted with the duties pertaining to the operation of steam boilers and steam machinery not exceeding 20 H. P.

All firemen who have charge of steam boilers, as to the regulation of feedwater and fuel, where the boilers are so situated as not at all times to be under the eye of the engineer in charge, are required to pass a third-class engineer's examination and procure the same kind of license. Engineers holding licenses of any of the preceding classes, and who are entrusted with the care and management of traction engines, or engines or boilers on wheels, other than locomotives, are required to pass an examination as to their competency to operate such class of machinery, and to procure a license known as a traction license. Such traction license shall not entitle the holder thereof to operate any other class of machinery.

All certificates of license to engineers of all classes shall be renewed yearly. The fee for renewal is $1 in all cases. The fee for the examination of applicants for engineers' licenses is $7.50 for first-class engineers, $5 for second-class engineers, $3 for third-class engineers, and $3 for traction engineers. Fees must be paid at the time of application for license. In case of the failure of any applicant to pass a successful examination, 90 da. must elapse before he can again be examined as an applicant for license in the class for which he was examined. But the inspector may grant to the applicant a lower grade of license than applied for on such examination.

Ohio.—In the state of Ohio, the license laws are administered by the chief examiner of steam engineers and a number of district examiners. The chief examiner is appointed by the governor of the state, and the district examiners are appointed by the chief examiner, with the approval of the governor.

The laws state that any person who desires to act as a steam engineer shall make application to the district examiner of steam engineers for a license so to act, on a blank furnished by the examiner, and shall successfully pass an examination on the construction and operation of steam boilers, steam engines, and steam pumps, and also hydraulics, under such rules and regulations as may be adopted by the chief examiner, which rules and regulations and standards of examination, however, shall be uniform throughout the state. If, on such examination, the applicant is found to be proficient in the prescribed subjects, a license shall be granted him to have charge of and operate stationary steam boilers and engines. It shall be unlawful for any person to operate a stationary steam boiler or engine of more than 30 H. P. without having been licensed to do so.   Boilers and engines under the jurisdiction of the United States and locomotive boilers and engines are excepted.

Licenses continue in force for 1 yr. from the date of their issue, unless something occurs to render the holder unfit to discharge the duties of steam engineer, in which case the license may be revoked.   Renewals of licenses are granted on application at the expiration of 1 yr. from the date of issue.   The fee for examination of applicants for license is $2, which must be paid at the time of application for examination; each renewal of license costs $2.

Pennsylvania.—In the state of Pennsylvania there are license laws relating to steam engineers and the inspection of steam boilers in cities of the second and third classes.   The administration of these laws is in the hands of the boiler inspectors of the several cities involved.   The councils of the cities provide for the creation of the office of boiler inspector.   No steam boiler or steam engine of over 10 H. P. may be operated by any person who is not over 21 yr. of age and who does not hold a license, except in the following cases: locomotive boilers used in transportation and steam boilers and engines carrying pressures of less than 15 lb. per sq. in.

Every person desiring authority to perform the duties of an engineer shall apply to the boiler inspector of such cities, who shall examine the applicant as to to his knowledge of steam machinery and his experience in operating it, and also the proofs he produces in support of his claim.   If the inspector is satisfied that the applicant's character, habits of life, knowledge and experience in the duties of an engineer are such as to authorize the belief that he is a suitable and safe person to be entrusted with the powers and duties of such station, he shall be granted a license on the payment of $3.   Licenses are to be renewed annually on the payment of $1 and within 10 da. after the expiration of date of such license.

Licenses are of two classes, namely, those entitling the holders to have charge of or to operate stationary steam boilers and steam engines only, and those entitling the holders to have charge of or to operate portable steam boilers and steam engines only.   Transferring from one grade to the other can be done only through a reexamination, but without cost to the licensee.

No person shall be eligible to examination for license unless he furnishes proof that he has been employed about a steam boiler or a steam engine for a period of not less than 2 yr., prior to the date of application, which proof must be certified by at least one employer and two licensed engineers. It shall be the duty of every licensed engineer when he vacates a position as engineer to notify the boiler inspector of such fact. Failure to do so is punishable by suspension of license for such a period of time as the boiler inspector may determine.

Tennessee.—In Tennessee, the license laws are administered by a board of inspectors appointed by the mayor or the president of cities having a population of 30,000 or over. The duties of the board are to examine into the qualifications of applicants for license to act as engineers of steam plants, The board holds sessions at least twice each month for the purpose of receiving applications for license. The laws state that the board shall grant certificates of license for 1 yr., to all applicants who, on examination, shall have the skill, experience, and habits of sobriety requisite to perform the duties of an engineer. Any owner or user of steam boilers of a capacity of not over 75 sq. ft. of heating surface and a pressure of not over 25 lb. per sq. in., and of all boilers of a pressure less than 15 lb. per sq. in. used for heating purposes only, may obtain a permit from the board to employ a careful and trustworthy person instead of a licensed engineer, such person to be recommended by two citizens, one of whom shall be a steam user or a licensed engineer. When boilers are used for engines run day and night, the owner or user of them may employ some trustworthy person in place of a licensed engineer, not exceeding 12 hr. at a time, under the instructions of a licensed engineer-in-charge.

In case an owner may be deprived of the services of a licensed engineer, he may put a careful and trustworthy person in charge for a time not exceeding 6 da. In places where there are steam boilers or steam-generating apparatus of over 10 H. P., and when such apparatus is in use, there must be employed at least one licensed engineer.

Applicants for engineer's license must make application on a blank furnished by the board for that purpose. Applicants must have experience of at least 2 yr. at mechanical or steam engineering, and must state their experience on the blanks. All applications must be signed by two citizens, one of whom must be a steam user or a licensed engineer, who shall go before the board and make oath that the statements set forth in the application are true. In taking charge of a plant, and when leaving a plant to assume charge of another, engineers must notify the board immediately in the first instance and 10 da. previous in the second instance. The fee for each license or renewal is $5.

ABSTRACTS OF CITY ENGINEERS' LICENSE ORDINANCES

Allegheny, Pa.—See Pennsylvania state law, on page 250.

Atlanta, Fulton County, Ga.—The license laws for Atlanta, which is in Fulton County, Ga., are, in part, as follows:

Any person desiring to be examined for a license to run and operate steam boilers and stationary engines in Fulton County shall make application in writing to the Board of Examiners of Engineers of Fulton County. Such application must be indorsed by three reputable citizens of this county, one of whom shall be a licensed engineer. The three citizens must certify to the good character and sober habits of the applicant and that he has had experience of not less than 1 yr. as an engineer, or experience of 1 yr. as an apprentice under a licensed engineer. No license shall be issued by the board of examiners to any one until after a full compliance with this rule.

A license shall be granted to only such engineers as, after a careful examination, the board of examiners shall be satisfied are competent to have full charge of the class of engines and boilers covered by the license issued to them.

The board of examiners shall classify all licenses in accordance with the character of the engines run by stationary engineers, from the plain slide-valve engine to that employing the most difficult and complicated machinery. The license must show on its face the character or class of same.

The board of examiners shall have the authority to issue licenses to assistant engineers, classifying them as above provided, and issue to the same only such license as may cover the character of the engine, which in their judgment such assistant engineer is competent to have charge of, and then only when the chief engineer in charge of the same engine has a full license covering the class to which it belongs.

Whenever any engineer or assistant engineer has received a license of one class, and desires to be licensed in a higher class, or, in the case of an assistant engineer, to be licensed as chief engineer, he may make application to the board of examiners for that purpose. After a reexamination the engineer or the assistant engineer may be granted such higher license if, in the opinion of said board of examiners, he is competent to take charge of such engine as may be covered thereby.

When any person shall, after written notification from the board of examiners, continue to run or operate any stationary engine or boiler in Fulton County without a license from them covering the class of engine that he has in charge, or whenever any person shall knowingly employ or cause to be employed any person to run or operate a stationary engine or boiler in Fulton County who has not been licensed by the board of examiners, such person shall be prosecuted by the board of examiners under the criminal laws of this state.

Each license issued by the board is issued and accepted subject to the rules of examiners.

The board of examiners shall receive a fee from each person examined for a license as either engineer or assistant engineer of $5 in each case, which fee shall be required to be paid prior to the examination of the applicant and whether the license is issued to him or not.

Baltimore, Md.—In the city of Baltimore there is a board of examining engineers appointed biennially by the governor of the state. The law states that this board shall have general supervision of all stationary engineers within the state of Maryland, except as hereinafter provided. It shall be the duty of the board to examine all engineers of the age of 21 yr. and upwards who shall apply to them for examination. Those who pass the examination and receive a certificate shall pay

18

the board the sum of $3 for each certificate so issued, and tor all renewals of all grades the sum of $1.50.

The certificates shall be of three grades. A certificate of the first grade will permit the holder to take charge of any plant of machinery; one of the second grade, to take charge of any plant of machinery from 1 to 500 H. P.; and one of the third grade, to take charge of any plant of machinery from 1 to 30 H. P. The said certificate shall run for the term of 1 yr. and shall be renewed annually.

All persons desiring to fill a position as a stationary engineer must make application to the board of examining engineers, with the following exceptions: Persons who are running engines and boilers in sparsely settled country places, where not more than twenty persons are engaged in work about such engines and boilers; engineers running country saw mills and grist mills, threshing machines, and other machinery of a similar character; marine engineers engaged in steamboats or any vessel run by steam; and persons engaged as locomotive engineers of any steam railway company.

The law also states that the headquarters of the board shall be in Baltimore and that it shall meet at least once in every week, and at a specified hour and day shall sit until all applicants shall be examined. If there are too many applicants to examine on the regular day for the purpose, the board shall continue its sessions until all applicants have been examined.

Buffalo, N. Y. —Every person within the city limits of the city of Buffalo in charge of or operating any steam engine or steam boiler (excepting persons operating locomotive steam engines or marine engines, or persons licensed as engineers by the authorities of the United States, or persons in charge of any steam engine or boiler in any of the public-school buildings, or any engineer while in the employ of the fire department of the city) shall appear in person before the examiner of stationary engineers for examination as to his qualifications as a stationary engineer, and if found qualified, shall be duly licensed, as the ordinance provides. But such persons who have charge of any steam boiler or steam engine in public-school buildings of the city shall be examined as to their qualifications to have charge of same.

No person shall be granted a license unless he be an actual resident of the city of Buffalo, and shall be a citizen, or shall have declared his intention to become a citizen, of the United States. All licenses must be renewed annually, and no person shall have charge of or operate more than one steam plant. A fee of $3 shall be collected by the examiner upon issuing a license, and $2 for each annual renewal.

The classification and grades are as follows: Chief engineer, first-class engineer, second-class engineer, and special engineer. The grades are according to the capacity and horsepower of a steam engine, steam boiler, or steam plant of which such engineers are found competent to take charge. Chief engineers are qualified to take charge of and operate any size of steam plant; first-class engineers, to take charge of any steam plant not exceeding 150 H. P.; second-class engineers, to take charge of and operate any size plant not exceeding 75 H. P.; and special engineers, to take charge of only a certain steam engine or boiler to be stated in the license, such steam engine or boiler not to exceed 10 H. P. and such license not to be used for a longer term than 1 yr.

Chicago, Ill.—In Chicago a board of examiners deals with the licensing of engineers. This board holds in quarters provided by the commissioner of Public Works, daily sessions for the purpose of examining and determining the qualifications of applicants for licenses for engineers. Every application for a license must be made on printed blanks furnished by the board of examiners; that for an engineer must be accompanied by a fee of $2, and that for a boiler or water tender must be accompanied by a fee of $1.

An applicant for an engineer's license must be a machinist or an engineer having a practice of at least 2 yr. in the management, operation, or construction of steam boilers and engines. An applicant for a boiler tender's license must be a person who has a thorough knowledge of the construction and management, and operation of steam boilers. Each applicant must state upon the blank the extent of his experience; must be at least 21 yr. of age, a citizen of the United States, or have declared his intention to become such; and must be of good character; all of which must be vouched for in writing by at least two citizens of Chicago, or may be verified under oath by the applicant when required by the board of examiners.

It shall be the duty of the board to see that each boiler plant in the city of Chi.ago shall have a licensed engineer, or boiler or water tender, or both as the case may be, in charge at all times when working under pressure; certificates must be displayed in a conspicuous place in the engine or boiler room. Each engine and boiler tender shall devote his entire time, while boilers are working under pressure, to the duties of the plant under his charge. Any person having charge of a steam boiler whose duty it is to keep up the water in such boiler shall be deemed a boiler or water tender within the meaning of the ordinance, but the provisions for the examination, licensing, and regulation of boiler or water tenders shall apply only to boiler or water tenders who ᴗre in charge of a steam boiler or boilers, that are detached from the engine room or so far removed therefrom or otherwise located as to render it difficult for the engineer in charge of the plant to give it or them his personal attention and supervision.

The following are exempt from the provisions of the ordinance: engineers in charge of locomotives and all boilers used for heating private dwellings, hothouses, conservatories, and other boilers carrying a pressure of not more than 10 lb. per sq. in. and the persons operating them.

Denver, Colo.—In Denver the mayor appoints a board of examiners consisting of the city boiler inspector and two practical engineers, whose duty it is to examine applicants for licenses as engineers and boiler or water tenders, in accordance with the rules and regulations of the ordinance, and to issue certificates of qualification. The law states that each certificate issued by the board shall expire 1 yr. from the date of issue, and that the board shall hold weekly sessions of such duration as may be deemed requisite for the purpose of examining and determining the qualifications of applicants for licenses as engineers or as boiler or water tenders.

Every application for a license must be made on printed blanks furnished by the board of examiners, and must set forth the name, age, and citizenship of the applicant and the extent of his experience. An application for an engineer's license must be accompanied by a fee of $2, and that for a boiler or water tender's license, by a fee of $1.

According to the ordinance, an applicant for an engineer's license shall be a machinist or engineer, having a practice of at least 2 yr. in the management, operation, or construction of steam engines and boilers, and an applicant for a boiler tender's license shall be a person who has a thorough knowledge of the construction, management, and operation of steam boilers. Each engineer and boiler or water tender so to be licensed shall be at least 21 yr. of age and of good character, all of which shall be vouched for in writing by at least two citizens of Denver, or shall be verified under oath by the applicant when required by the board of examiners.

All such licenses may be renewed from year to year upon payment of the license fee before specified and without further examination, unless the applicant applies for a different class or grade of license.

Engineers in charge of locomotives and engineers or boiler or water tenders in charge of boilers carrying a steam presssure of not more than 10 lb. per sq. in. are exempt from the provisions of the ordinance.

Detroit, Mich.—In Detroit the city boiler inspector examines candidates as to their fitness to perform the duties of the stationary engineer. Any person claiming to be qualified to operate a boiler must apply in writing on blanks provided for the purpose for a license. The inspector shall examine him and consider the proof offered in support of his claims. If the inspector is satisfied that the applicant's knowledge, experience, and character render him competent to handle boilers with safety, he shall issue a license certificate to that effect, designating the class in which the engineer is authorized to operate.

There shall be three grades of engineers' license. First-class engineers' licenses shall be unlimited as to the number of boilers and pressure, and shall be granted to any citizen having an experience of 5 yr. in the care of steam boilers, provided he can pass a satisfactory examination. Second-class licenses shall be limited to 75 H. P. and shall be granted to any citizen having an experience of 3 yr. in the care of steam boilers, provided he can pass a satisfactory examination.

Third class licenses shall be limited to 25 H. P. and shall be granted to any citizen having experience in firing steam boilers for 2 yr., provided he can pass a satisfactory examination.

If any person makes application for a certain class of license and fails to procure it for any cause, the inspector can assign him to the class to which his examination entitles him, and he cannot make application again for a period of less than 3 mo. Any person having a second-class or a third-class license can act as assistant to a first-class engineer.

The fee for licenses shall be $1, payable before examination, but one-half of the amount paid shall be refunded if the license is refused.   The provisions of the ordinance do not apply to locomotive boilers used on railroads, boilers under the jurisdiction of the United States, boilers in the fire department of the city, and boilers used in private residences for heating purposes.

Elgin, Ill.—The engineers' license laws for Elgin, Ill., are, in part, as follows:

There shall be and there is hereby authorized to be appointed by the mayor, by and with the consent of the city council, a board of examining engineers.   The said board shall consist of three practical engineers.   The duties of the board shall be to examine into the qualifications of applicants for engineer's license; to license those found qualified, and, for cause, to suspend or revoke the same.

The examination may be written or oral and shall be entirely practical.   Applications for license shall be made on a printed blank furnished by a board of examiners and shall set torth the name, age, citizenship, residence, experience, etc., of the applicant, together with the location of the plant for which a license is desired.   Each application for license shall be accompanied with $1, which fee shall be returned in case of failure to secure a license.

The board of examiners may suspend the license of any engineer for carelessly permitting the water to get too low and burning the boiler; for carrying the steam pressure higher than allowed by law; for absence from his post of duty; for neglect, incapacity, or intoxication while on duty; provided that no license be suspended or revoked without first giving the accused person an opportunity to be heard in his own defense.

The license will be suspended not to exceed 30 da. for the first offense, not to exceed 90 da. for the second offense, and not to exceed 6 mo. for an offense after that.

Engineers' licenses will be good for 1 yr. from date of issue and must be renewed annually. Engineers who change from one plant to another must secure permission from the board of examiners, who will either authorize the change or reexamine the applicant, at their discretion, without additional cost to applicant. Engineers' licenses must be framed under glass and kept in a conspicuous place in engine room or boiler room.

All applicants for an engineer's license shall have a practical experience of at least 1 yr. in the management of steam engines and boilers. Each applicant shall be at least 20 yr. of age (except that by unanimous consent of the board a younger man may be licensed) and must be of temperate habits and good character, all of which must be vouched for in writing by two freeholders of Elgin, or verified under oath by the applicant when required by the examiners.

It shall be unlawful for any unlicensed person to take charge of or operate any steam power or boiler plant within the city of Elgin, excepting engineers in charge of locomotives, steam-road carriages, and those in charge of boilers carrying a steam pressure of less than 15 lb. per sq. in. Any person who shall take charge of or operate any steam power or boiler plant for a longer time than to the next meeting of the board of examiners, shall be guilty of a misdeameanor and subject to a fine of not more than $20 nor less than $5. Any person, company, or corporation controlling any steam power or boiler plant who shall authorize or permit any person, without proper and valid license, to take charge of or operate the same for a longer period than above stated, shall be guilty of a misdemeanor and subject to a fine of not more than $50, nor less than $20, for each offense, and each day's violation of this ordinance shall constitute a separate offense.

Any engineer whose license has expired for a period of 10 da. must give satisfactory explanation to the board of examining engineers why such a condition exists, and any engineer whose license has expired for a period of 1 yr. or more shall be subjected to a reexamination.

Any engineer or water tender changing his position from one plant to another shall within 5 da. have his license transferred to the plant he is to operate.

Goshen, Ind.—In the city of Goshen, Ind., there is appointed by the mayor, with the consent of the common council, an examining engineer who has charge of matters pertaining to the licensing of stationary engineers.

Any person desiring to act as an engineer shall make application to the examiner, upon blanks furnished by the examiner, and if, upon examination, the applicant is found to be trustworthy and competent, a license shall be granted to said applicant; such license shall continue in force for 1 yr., unless for cause it may be revoked.

License shall be granted according to competency of the applicant, and shall be divided into classes as follows: Class 1, the engineer's license for which shall be unlimited as to horsepower; class 2, the license for which shall be limited to 150 H. P.; and class 3, the license for which shall be limited to 50 H. P. Engineers holding second- or third-class licenses may act as assistant engineers in a plant of any capacity, provided the man in charge holds a first-class license.

The fee for examination of applicants shall be $1, to be paid at the time of the application for examination, and $1 for each renewal of license. Locomotive boilers and engines, boilers of private residences, or boilers of less than 8 H. P. do not require a licensed engineer to operate them.

Hoboken, N. J.—License laws in Hoboken, N. J. are administered and enforced by a board of examining engineers, consisting of five members appointed by the council. The mayor and the council of the city of Hoboken do ordain as follows:

No person shall be the engineer of, or shall have charge of, or operate any steam boiler, or steam engine in the city of Hoboken, for a period exceeding 2 da., who shall not have a license certificate or shall have made application for the same authorizing him to have charge of, or to operate such engine, or boiler, from the board of examining engineers, and no such license shall be granted unless the applicant therefor be a citizen of the United States.

Before any person shall be employed as an engineer of any such steam boiler or engine, or shall have charge of or operate any such boiler or engine, he shall make a written application on blanks furnished by the city to said board of examining engineers, for the license heretofore mentioned, which application shall be accompanied by references as to the character and ability of the applicant, and the filing of such references with said board shall be considered as a compliance with the provisions of this ordinance for 30 da. thereafter or until the said application shall have been passed upon by said board, and said applicant after the filing of said reference shall have the right to operate and have charge of any such engine, boiler or plant until his application shall have been passed upon by said board.

Every person who shall satisfy said board of examining engineers that he is a safe and competent person to operate and have charge of such steam boiler, engine or plant specified in his application, shall, upon payment of $2 to the city clerk as his fee, receive a license permitting him to operate the same for 1 yr., unless such license shall be sooner revoked. For an annual renewal of such license the licensee shall pay to the city clerk as his fee the sum of $1 therefor; additional hearing shall not be required, unless in the judgment of the said board the same may be necessary. Such licenses must be framed and hung in a conspicuous place in the plant, or upon, or near the engine or boiler, in charge of such licensee.

Said board may at any time after proper hearing revoke any license issued on account of inebriety, incompetency, or negligence of the holder of any such license, or for any other good cause, and no license shall be issued to any licensee whose license shall have been revoked for a period of 6 mo., after which the license revoked may be renewed, if in the judgment of the board the cause of its revocation no longer exists.

If said board shall refuse to grant to any applicant a license, no license shall be issued to him for the next 6 mo. following the refusal of said application, but after said period said applicant may make another application, and, if qualified, may be granted a license.

This ordinance shall not apply to locomotive engineers or to engineers on steam vessels coming under the jurisdiction of the United States Board of Supervising Inspectors, nor shall it apply to boilers in private residences for heating purposes, unless, in the opinion of said board, such boiler is so equipped and run as to endanger public safety unless operated by a licensed engineer.

Any engineer in charge of any steam engine or boiler who shall abandon it while in operation without leaving a person in charge of same, who shall, in the opinion of the board of examining engineers, be competent to take charge of same, shall be fined the sum of $10.

Huntington, W. Va.—License laws in the city of Huntington, W. Va., are administered and enforced by a board of engineers consisting of four members. Candidates for examination must have had at least 1 yr. of experience in operating steam engines and boilers. The law states that it shall be the duty of the board of engineers, upon the payment of $3 by the applicant, to examine persons touching their qualifications as engineers, who desire to act as engineers and take charge of steam boilers, engines, and pumps. If the applicant passes a satisfactory examination, the board shall grant and issue to him a certificate of qualification. If the applicant fails to pass a satisfactory examination, he shall not be allowed to apply again for certificate for 2 mo. thereafter. All certificates granted shall be in force for 1 yr, from the date thereof and no longer, and any person holding a certificate from the board may have the same renewed at its expiration for a period of 1 yr. by the applicant for such renewal paying the sum of $2, provided, however, that the person applying for such renewal is entitled thereto, and such application for renewal is made on or before the last regular meeting of the board before the expiration of the applicant's certificate. Unless the above provision is complied with, the board may, at its discretion, order a new examination.

The board shall have the right to adopt rules and regulations as they deem necessary and proper, not inconsistent with this ordinance. The full board, by an unanimous vote, shall have power and may revoke any engineer's certificate upon cause being shown therefor.

It shall be the duty of each engineer holding a certificate from the board to display the same in some prominent place near the boiler or boilers in his charge.   The board may revoke the certificate of any engineer who shall fail or refuse to comply with this section.

It shall be unlawful for any person to operate or cause to be operated any steam boiler used to furnish steam at a pressure to exceed 25 lb. per sq. in., unless there be in charge of such boiler an experienced person having a certificate from the board of engineers, and any person found in charge and operating a boiler not having a certificate from the board of engineers shall be deemed guilty of a misdemeanor and, upon conviction thereof, shall be fined in the sum of not less than $5, nor more than $50 for each offense; provided, however, that any owner or user of any boiler in use which shall for any cause be deprived of the service of a person holding such certificate may procure an experienced and careful person to take charge of such boiler for a period not to exceed 10 da.

Jersey City, N. J.—Engineers' license laws are administered by a board of examiners in Jersey City, N. J.   The laws, in part, are as follows:

No person shall be the engineer of, or shall have charge of or operate any steam boiler or steam engine, in the city of Jersey City, for a period exceeding 1 wk., who shall not have a license certificate authorizing him to have charge of or operate such engine or boiler, from the board of examiners.

Before any person shall be employed as an engineer of any such steam boiler or engine, or shall have charge of or operate any such boiler or engine, he shall make a written application to the board of examiners for the license, and shall specify in such application the particular engine, boiler, or plant that he desires to operate, or have charge of, which application shall be accompanied by references as to his character and ability, and the filing of such reference with such said board shall be considered as a compliance with the provisions of this ordinance for 30 da. thereafter, or until his said application shall have been passed upon by said board, and said applicant, after the filing of said references, shall have the right to operate until his application shall have been passed upon by said board.

Every person who shall satisfy said board of examiners that he is a safe and competent person to operate and have charge of the steam plant, boiler, or engine specified in his application, shall, on payment of $1 to the city clerk, for the benefit of the city, receive a license permitting him to operate the same for 1 yr., unless sooner revoked. Said license shall apply only to the plant, boiler, or engine for which it is issued, and before taking charge of another plant the licensee shall apply for another license for such other plant, for which other license, if the application be made within a year, no charge shall be made. For annual renewals of such licenses a fee of $1 shall be paid. For the renewals above mentioned, no additional hearing shall be required, unless in the judgment of said board it shall be necessary. Said licenses must be framed and hung in a conspicuous place in the plant, or upon or near the engine for which is is issued.

If the board shall refuse to grant to any applicant a license, no license shall be issued to him for the next 6 mo. following the refusal of his application, but after said period said applicant may make another application, and if found qualified, may be granted a license.

Whenever said board shall refuse to grant any application, or shall revoke any license, they shall give immediate notice of such refusal or revocation to the applicant or licensee, and such applicant or licensee may appeal from the decision of such board to the board of aldermen, in which case said applicant or licensee shall file his appeal with the board of aldermen within 10 da. after receiving notice of the decision of said board, and the board of aldermen may confirm or reverse the decision of said board, and issue such license.

The ordinance shall not apply to railway locomotives, nor to engineers employed thereon, nor to steam vessels coming under the jurisdiction of the United States Board of Supervising Inspectors, when employed upon the vessel to which said license applies. Nor shall it apply to boilers in private residences or buildings for heating purposes, unless, in the opinion of said board, such boiler is so equipped and run as to endanger public safety unless operated by a licensed engineer.

Kansas City, Mo.—In Kansas City, Mo., a board of engineers appointed by the mayor, with the consent of the council, administers the license laws. The board convenes for business once each month, to examine into the qualifications of applicants for engineers' licenses. According to the laws, the board shall grant certificates of license, charging $5 to each applicant for the first certificate, $3 of which is to be deposited at once. Each applicant is to be allowed three trials, and if he then fails to pass a satisfactory examination, the applicant shall forfeit the money deposited (namely, $3) with the clerk of the board. But if the applicant has the capacity, skill, experience, and habits of sobriety requisite to perform the duties of an engineer, and shall pass the examination, the board shall grant him a license for the term of 1 yr. upon the payment of an additional $2. Any person so qualified shall not be refused a license.

Renewal of license will be granted to applicants, upon payment of $2.50, if renewed on or before the next regular meeting of the board of engineers after its expiration. All engineers, engines, and boilers of the fire department of Kansas City, the locomotive boilers used on railroads, and steam boilers supplied with water automatically, when used only for heating dwelling houses and not carrying a pressure over 10 lb. per sq. in., are exempt from the provisions of this ordinance.

Every applicant for a license who fails to pass the examination of the board is required to wait 4 wk. before again making application for license, and the board shall then give him another examination. Applicants failing to pass the examination after the third trial shall not be permitted to appear again before the board for 6 mo.

Every engineer licensed by the board and under control of the board is required to notify the boiler inspector—who is a member of the examining board referred to—when he accepts employment, and within 3 da. thereafter, ·the name of his employer, and the location of the boilers in his charge. Any engineer who shall neglect or refuse to comply with this rule shall be deemed guilty of a misdemeanor. Engineers shall report semiannually to the boiler inspector, during the first 3 da. of the months of January and July, the condition of the boilers, pumps, and connections under their charge. Failure to comply with the rule shall be deemed a misdemeanor.

Lincoln, Neb.—License laws in Lincoln, Neb., are administered by a board of engineers, which consists of three members appointed by the mayor with the consent of the city council.

The law states that it shall be the duty of the board of engineers, on the payment of $3 by the applicant, to examine persons, touching their qualifications as engineers, who desire to act as engineers and take charge of steam boilers. All certificates granted shall be in force for 1 yr. from the date thereof and no longer; and at the end of 1 yr. certificates may be renewed for another year by the applicant paying the sum of $2, provided the person applying for such renewal is entitled thereto and such application is made on or before the last regular meeting of the board, before the expiration of the applicant's certificate.

. It shall be the duty of every person holding a certificate from the board to make a semiannual report to the board, during the months of January and July of each year, of the condition of every boiler, pump, and connection under his charge. Failure to make such report is sufficient cause for the board to revoke the certificate of the person involved. Heating apparatus in private dwellings are exempt from inspection, as provided by the ordinance in this city; also those boilers kept insured in any reputable and legitimate insurance company requiring inspection.

Four grades of certificates, namely, first-, second-, and third-grade certificates and a certificate specified low pressure, are issued by the board. First-grade certificates are granted to applicants who, on examination, are found qualified to take charge of and operate any steam plant; second-grade certificates are granted to applicants who, on examination, are found qualified to take charge of and operate any steam plant up to 75 H. P. only; and third-grade certificates are granted to applicants who, on examination, are found qualified to take charge of and operate any steam plant up to 25 H. P. only. Certificates specified low pressure are granted to applicants who, on examination, are found qualified only to take charge of and operate low-pressure boilers for heating purposes.

Los Angeles, **Cal.**—The license laws in Los Angeles are administered by the city boiler inspector and an assistant inspector appointed by the city council, together with a board of examining engineers, of which there are three in number.

The board, so the law states, shall hold one meeting on the first and third Wednesday in each month for the purpose of examining applicants for engineer's license, and shall hold the meeting on the second and fourth Tuesday of each month for the purpose of examining applicants for elevator license. The board shall make a careful and thorough examination as to the qualifications of all applicants for engineer's license, and shall grant certificates of license to all persons found qualified; it shall charge and collect from each applicant for a chief, or first-class, license, the sum of $5, and from each applicant for a second or a third-class license, the sum of $3. Such licenses are good for the term of 1 yr., unless revoked for cause.

In case any owner or user of any boiler shall for any cause be deprived of the services of a licensed engineer, he must notify the boiler inspector at once, and may place an experienced person in charge, for a time not beyond the date of the next regular meeting of the board of engineers. When boilers are used and engines run night and day, the owner or user of them must employ at least two licensed engineers, who shall stand watch alternately. No person shall use or operate any steam boiler or steam-generating apparatus in the city of Los Angeles without obtaining a certificate of license as provided for; this applies to apparatus of over 5 H. P.

Applicants for license who fail to pass the examination of the board of engineers, shall be required to wait for 4 wk. before making another application. Applicants who fail to pass after a third trial shall not be permitted to again appear before the board for 6 mo. An engineer must notify the boiler inspector of any employment that he may enter into as an engineer, and within 3 da. after, the name of his employer and the location of the plant in his charge. Engineers must also report semiannually to the boiler inspector, during the first 3 da. of the months of January and July of each year, the condition of the boiler or other apparatus and their connections under or in his charge.

Applicants for renewal of licenses shall pay to the board of engineers the sum of $1 for each yearly renewal.

Any person violating any of the provisions of this ordinance shall be deemed guilty of a misdemeanor, and upon conviction shall be punished.

Memphis, Tenn.—In the city of Memphis, Tenn., there is a board of examiners appointed by the legislative council. This board consists of the city boiler inspector and four practical steam engineers, and is created for the purpose of examining and licensing engineers having charge of or operating boilers and steam engines in the city of Memphis.

The laws of this city state that the board of examiners shall hold at least two sessions each month, on the first and third Mondays, for the purpose of receiving and acting on applications for license. The board shall grant certificates of license for 1 yr. to applicants who, on examination, shall have the skill, experience, and habits requisite to perform the duties of an engineer. Licenses shall be renewed without examination upon payment of the required fee. Applicants for license must not be less than 21 yr. of age, and must be citizens of the United States or have declared their intention to become citizens. Applications must be made upon blanks furnished by the board for such purpose. Applicants must have an experience of at least 3 yr. at mechanical or steam engineering, and so state it on the blanks. All applications must be signed by two citizens of the United States, one of whom must be an engineer or steam user, and both of whom must make affidavit before an officer, qualified to administer an oath, that the statements set forth in such applications are true. When an applicant fails to pass an examination three times, he will not be eligible to take another examination until 60 da. has passed from the time of his last appearance before the board.

Licenses shall be of three grades, namely, first, second, and third. A first-grade license shall entitle its rightful holder to operate or to have charge of steam plants of unlimited capacity as to horsepower of boilers. A second-grade license shall entitle its rightful holder to assist a first-grade-license engineer in any steam plant where such services are required under the instructions of first-grade engineer in charge of steam plant;

or he may have charge of and operate steam plants limited to 75 H. P. of boilers. A third-grade license shall entitle its rightful holder to assist first- or second-grade engineers where such services are required, under the instructions of engineer in charge of steam plant; or he may have charge of or operate steam plants limited to 25 H. P. of boiler.

The fee for each license or yearly renewal shall be $2. The provisions of this ordinance shall apply to all steam plants or boilers operated within the city limits, except locomotive boilers used on railroads. If any owner or user of any boiler or boilers or steam-generating apparatus is deprived of the services of the licensed engineer or engineers employed by him, for any reason over which he has not control, he may employ an unlicensed engineer for 30 da., in which time he must secure a licensed engineer.

No person shall receive a license who is not able to determine the weight necessary to be placed on the lever of a safety valve (the various data to work the problem being given) to withstand any given pressure of steam in the boiler. Applicants must also be able to figure and determine the strain brought on the braces of a boiler with a given pressure of steam, the position and distance apart being known. They must be able to figure and determine the safe working pressure of a boiler, such knowledge to be determined by an examination in writing. No license, except third grade, shall be granted to any engineer who does not possess the foregoing qualifications. No license shall be granted to any engineer who cannot read and write and does not understand the plain rules of arithmetic. The examination questions asked an applicant for license shall be practical ones pertaining to boiler and engine care and management; correct answers to 80% of such questions shall qualify the applicant to receive his license.

Milwaukee, **Wis.**—The engineers' license laws in the city of Milwaukee are administered and enforced by a board of examining engineers, consisting of two persons.

No person may operate or have control of any stationary or portable steam boiler, engine, or any portion of a steam plant, over 10 H. P., when working under pressure, except a duly licensed engineer.

19

The board of examiners shall hold daily sessions for the pur-pose of examining and determining the qualifications of appli-cants for licenses for engineers and for persons having charge of steam boilers or engines, as provided in the ordinance.

All persons desiring to perform the duties of a stationary or portable-boiler engineer shall make application therefor to the board of examiners, and shall present therewith a receipt from the city treasurer for a fee of $3. The board shall have the power to examine applicants, to grant licenses, and to revoke or suspend the same for cause. Applications must be made on printed blanks furnished by the board.

Licenses shall be in force for 1 yr. from date of issuance, and at the expiration of 1 yr. and on the payment of $1 licenses may be renewed without further examination. An applicant for engineer's license must be a machinist, engineer, oiler, or fire-man having an experience of at least 2 yr. in the management, operation, or construction of steam boilers and engines. Each applicant must state the extent of his experience, be at least 21 yr. of age, and of good character, all of which must be vouched for in writing by at least two citizens of the city of Milwaukee.

If the applicant, on examination, be found qualified as an engineer of a stationary or a portable steam boiler or engine, he shall be granted a license according to class, as provided for. Licenses so granted shall be graded into three classes: (1) Per-sons holding first-class licenses may take charge of and operate any steam- or motive-power plant; (2) persons holding second-class licenses may take charge of and operate any steam- or motive power plant not exceeding 300 H. P.; (3) persons hold-ing third-class licenses may take charge of and operate an steam- or motive-power plant not exceeding 75 H. P. Twelve sq. ft. of boiler heating surface shall be equivalent to 1 H. P.

The holder of a second-class license may act as an assistant engineer to an engineer in charge of a plant under a first-class license; and the holder of a third-class license may act as an assistant engineer to an engineer in charge of a plant under a second-class license.

It shall be unlawful to carry a higher pressure of steam than that fixed by the board. Engineers' licenses must be displayed under glass, in a conspicuous place in the boiler or engine room. It shall be the duty of every licensed engineer to report to the board of examiners any defects in any steam boiler, engine, or appurtenance belonging thereto under his charge.

This ordinance does not apply to engineers in charge of locomotives or to those in charge of engines or steam-boiler machinery under the civil service of the city, county, state, or federal government, or to engines or steam boilers used for heating private dwellings, and other engines or steam-boilers carrying a pressure of less than 15 lb. per sq. in.

Mobile, Ala.—A board of examiners of engineers, consisting of three members and elected by the general council, administers the license laws in the city of Mobile. The board must meet at least twice each month for the despatch of such business as may come before it.

According to the laws of this city, no person shall be entrusted with, or have charge of, the management or operation of any steam boiler having more than 200 sq. ft. of heating surface, or carrying a pressure greater than 20 lb. per sq. in., until such person has been duly examined by the board of examining engineers.

Whenever any person makes application, or is called before the board for examination, he shall be thoroughly examined as to his competency to manage and operate steam boilers, feed pumps, and injectors. If the applicant is found competent by the board, a license certificate shall be issued to him. Such license shall be in force for 1 yr. from the date of issuance. A new license may be issued to the said holder at the expiration of each year thereafter without further examination, as long as the holder continues in operating steam-engineering occupation. Should an engineer abandon his occupation as an operator for more than 1 yr., his license certificate shall become null and void, and it shall be necessary for said person to be reexamined by the board to obtain a new license certificate, before again undertaking to manage and operate steam boilers, feed pumps, and injectors.

Engineers are required to report in writing to the board of examining engineers, during the months of January and June of each year, the condition of the boiler or boilers under their charge; this also applies to those engineers who have charge of elevators. All reports mÙst contain full detailed and accurate information, and must be made on blank forms to be obtained from the board.

For each original certificate of license issued by the board, a sum of $2 shall be paid by the person to whom the license is issued; and for each renewal of license the sum of $1 shall be paid. These fee shall be collected by the board of examining engineers.

New Haven, Conn.—A board of examiners, appointed by the mayor and consisting of three members, administers the license laws in the city of New Haven.

No person shall be engineer of, or shall have charge of, or operate any steam boiler or steam engine in the city of New Haven, for a period exceeding 1 wk. who shall not have a license certificate from the board of examiners, authorizing him to have charge of and operate such boiler or engine. This ordinance shall not apply to railway locomotives nor engineers employed thereon, nor to steam vessels coming under the jurisdiction of the United States inspectors, nor to boilers in private residences or buildings for heating purposes, unless, in the opinion of the board, such boiler is so equipped and run as to endanger public safety, unless operated by a licensed engineer.

Before any person shall be employed as an engineer of any steam boiler or engine he shall make written application to the board for a license. He shall specify in the application the particular engine, boiler, or plant that he desires to operate or have charge of, and the application shall be accompanied by references as to his character and ability. After filing such references, he may operate a plant for 30 da. thereafter or until his application shall have been passed upon by the board.

Every person who shall satisfy the board of examiners that he is a safe and competent person to operate and have charge of the steam plant, boiler, or engine specified in his application, shall on the payment of $1 receive a license permitting him to operate the same for 1 yr. unless sooner revoked for cause. Said license shall apply only to the plant for which it is issued. Before taking charge of another plant, the licensee shall apply for another license for such other plant, for which other license if application be made within a year, no charge shall be made. For annual renewals of such licenses, a fee of $1 shall be paid.

The board may at any time revoke any license for cause, such as incompetency, neglect, or inebriety. In case a license is revoked, application may be made for a new license 6 mo. after, and on examination, provided the applicant is found qualified, the board may grant a license to him.

New York, N. Y. —The engineer's license laws for the city of New York are, in part, as follows:

No owner, or agent of such owner, or lessee of any steam boiler to generate steam shall employ any person as engineer or to operate such boiler unless such person shall first obtain a certificate as to qualification therefor from a board of practical engineers detailed as such by the police department, such certificate to be countersigned by the officer in command of the sanitary company of the police department of the city of New York. In order to be qualified to be examined for and to receive such certificate of qualification as an engineer, a person must comply, to the satisfaction of said board, with the following requirements:

1. He must be a citizen of the United States and over 21 yr. of age.

2. He must, on his first application for examination, fill out, in his own handwriting, a blank application to be prepared and supplied by the said board of examiners, and which shall contain the name, age, and place of residence of the applicant, the place or places where employed and the nature of his employment for 5 yr. prior to the date of his application, and a statement that he is a citizen of the United States. The application shall be verified by him, and shall, after the verification, contain a certificate signed by three engineers, employed in New York City and registered on the books of said board of examiners as engineers working at their trade, certifying that the statements contained in such application are true. Such application shall be filed with said board.

3. The following persons, who have first complied with the provisions of subdivisions one and two of this section, and no

other persons may make application to be examined for a license to act as engineer:

(*a*) Any person who has been employed as a fireman, as an oiler, or a as general assistant under the instructions of a licensed engineer in any building or buildings in the city of New York, for a period of not less than 5 yr.

(*b*) Any person who has served as a fireman, oiler, or general assistant to the engineer on any steamship, steamboat, or on any locomotive engine for the period of 5 yr. and shall have been employed for 2 yr. under a licensed engineer in a building in the city of New York.

(*c*) Any person who has learned the trade of machinist, or boilermaker, or steam fitter and worked at such trade for 3 yr. exclusive of time served as apprentice, or while learning such trade, and also any person who has graduated as a mechanical engineer from a duly established school of technology, after such person has had an experience of 2 yr. in the engineering department of any building or buildings in charge of a licensed engineer, in the city of New York.

(*d*) Any person who holds a certificate as engineer issued to him by any duly qualified board of examining engineers existing pursuant to law in any state or territory of the United States and who shall file with his application a copy of such certificate and an affidavit that he is the identical person to whom said certificate was issued. If the board of examiners of engineers shall determine the applicant has complied with the requirements of this section, he shall be examined as to his qualifications to have charge of, and operate steam boilers and steam engines in the city of New York, and if found qualified said board shall issue to him a certificate of the third class. After the applicant has worked for a period of 2 yr. under his certificate of the third class, he may be again examined by said board for a certificate of the second class, and if found worthy the said board may issue to him such certificate of the second class. After he has worked for a period of 1 yr. under said certificate of the second class he may be examined for a certificate of the first class, and when it shall be made to appear to the satisfaction of said board of examiners that the applicant for either of said grades lacks mechanical skill, is a person of bad habits, or is addicted to the use of intoxicating beverages, he shall not be entitled to receive such grades of license and shall not be reexamined for the same until after the expiration of 1 yr.

Every owner or lessee, or the agent of the owner or lessee, of any steam boiler, steam generator, or steam engine aforesaid, and every person acting for such owner or agent is hereby forbidden to delegate or transfer to any person or persons other than the licensed engineer the responsibility and liability of keeping and maintaining in good order and condition any such steam boiler, steam generator, or steam engine, nor shall any such owner, lessee, or agent enter into a contract for the operation or management of a steam boiler, steam generator, or steam engine, whereby said owner, lessee, or agent shall be relieved of the responsibility or liability for injury that may be caused to person or property by such steam boiler, steam generator, or steam engine. Every engineer holding a certificate of qualification from said board of examiners shall be responsible to the owner, lessee, or agent employing him for the good care, repair, good order and management of the steam boiler, steam generator, or steam engine in charge of, or run, or operated by such engineer.

Niagara Falls, N. Y.—A board of examiners of stationary engineers, consisting of three members, administers the license laws in the city of Niagara Falls. Any person desiring a license to act as a stationary engineer or fireman in this city, may file with the city clerk an application, together with the required fee for a license, provided one is granted, which fee is returned to the applicant in case a license is not granted to him. The application is to be submitted to the board of examiners by the city clerk, when such board shall examine the applicant at a time and place named by the board.

It shall be unlawful for any person to have charge of or operate any steam plant, steam engine, or steam boiler within this city, excepting locomotive steam engines, or marine engines, or engines used by the fire department of the city of Niagara Falls, without having procured a license from the board of examiners.

The board of examiners shall issue licenses of the following classes:

(*a*) A license as chief engineer to any person found qualified to take charge of and operate any steam plant.

(*b*) A license as first-class engineer to any person found qualified to take charge of a steam plant or engine not exceeding 300 H. P.

(*c*) A license as first-class fireman to any person found qualified to take charge of a steam boiler not to exceed 300 H. P.

(*d*) A license as second-class engineer to any person found qualified to take charge of and operate any steam plant or engine not exceeding 75 H. P.

(*e*) A license as second-class fireman to any person found qualified to take charge of and operate any steam boiler not exceeding 75 H. P.

(*f*) A license as special engineer or fireman to any person found qualified to take charge of and operate a certain steam engine or steam boiler of a certain horsepower capacity, to be designated in the license.

(*g*) A license to the owner of a steam engine or boiler using not to exceed 10 H. P., and used by him in his own business.

Licenses are granted for 1 yr. and the fees for them are as follows: For any license, save the license provided for in subdivision (*g*), $3. For a license issued under subdivision (*g*), $1. Licenses are to be renewed annually and the fee for each renewal is $2.

It shall be unlawful for any person to take charge of a steam engine, a steam boiler, or a steam plant of greater capacity and horsepower than that authorized by his license; and it shall also be unlawful for any person to have charge of or operate more than one plant at the same time. It shall be unlawful for any engineer or fireman to be absent from the steam boiler or the steam engine operated by him for more than 20 min. nor to be farther distant from such plant than 100 ft., while working under pressure.

It shall not be necessary to procure a license to take charge of or to operate a steam boiler used in private residences in cases where the water returns automatically to the boiler and the pressure does not exceed 10 lb. per sq. in.

Omaha, Neb.—In the city of Omaha a board of engineers administers and enforces the engineers' license laws. This board consists of the city boiler inspector and two practical and mechanical engineers, all of whom are appointed by the mayor, with the consent of the city council. The board convenes once in each month to examine into the qualifications of applicants for engineers' certificates.

Every applicant for a certificate who fails to pass the examination of the board is required to wait 3 mo. before again making application for a certificate. At the expiration of that time, the board will give him another examination. Applicants for examination will be furnished with a blank for the purpose by the board. Applicants must have an experience of at least 2 yr. at mechanical or steam engineering, and must write and state their experience in the blank. All applications must be signed by two citizens, one of whom must be a steam user or engineer. Each applicant is required to make oath before the board as to the truth of the statements in the blank. Engineers holding certificates granted by the board are required to notify the board when accepting or leaving employment; also to state the name of the new employer and the location of the boiler in their charge. This must be done immediately.

Engineers holding certificates granted by the board are required to make out a report as to the condition of all boilers and apparatus under their charge and to send it to the board; this must be done during the first 10 da. of January and July of each year.

The board shall issue three grades of certificates, as follows:— First-grade certificates shall entitle the holder to take charge of and run any plant; second-grade certificates shall entitle the holder to take charge of and run any steam plant under 100 H. P., and third-grade certificates shall entitle the holder to take charge of and run any steam plant under 50 H. P. The board of engineers may grant to persons operating low-pressure, gravity steam-heating plants, carrying a pressure not to exceed 20 lb. per sq. in., a special third-grade certificate to be valid for one particular specified plant and no other.

A fee of $5 shall be charged for each examination of an engineer for license by the board. Applicants must be at least 21 yr. of age. Certificates are valid for 1 yr. and no longer, but may be renewed each year upon the payment of $1 to the city treasurer and then presenting his receipt for the same to the city boiler inspector; where boilers are used and engines run night and day, the owners or users of steam power must employ two certified engineers, who may stand watch alternately.

Peoria, Ill.—A board of engineers consisting of three members, including the city boiler inspector, appointed by the Mayor and approved by the council administers and enforces the license laws in the city of Peoria.

According to the laws, the board shall hold sessions twice each month for the purpose of examining and determining the qualifications of applicants for engineers' licenses. The certificates of license shall be of two grades—first and second—and the requirements for each grade shall be determined by the board of examiners.

Every application for a license must be made on the printed blank furnished by the board for that purpose; that for an engineer must be accompanied by a fee of $5, and that for a boiler or a water tender must be accompanied by a fee of $2. Licenses are in force for 1 yr. only from date of issue. Reissues are made upon the payment of a fee of $2 for an engineer's license and a fee of $1 for a boiler tender's license.

An applicant for license must have had a practice of at least 3 yr. under the supervision of a practical engineer. An applicant for a boiler tender's license must be a person who has a thorough knowledge of the construction, management, and operation of steam boilers. Each applicant must state upon the blank the extent of his experience, he must be at least 21 yr. of age and a citizen of the United States, or have declared his intention to become such, and he must be of good character. All of this information must be vouched for in writing by at least three first-class engineers of Peoria, or it may be verified under oath by an applicant when required by the board.

It shall be the duty of the board of examiners to see that each boiler plant in the city shall have a licensed engineer or boiler or water tender, or both, in charge at all times when working under pressure. Any person who has charge of a steam boiler, whose duty it is to keep up the water in such boiler, shall be deemed a boiler or water tender within the meaning of this ordinance, but the provisions for the examining and licensing of a boiler or water tender shall apply only to boiler or water tenders who are in charge of a steam boiler or boilers that are detached from the engine room or so far removed therefrom as to render it difficult for the engineer in charge of the plant to give it or them his personal attention or supervision.

Every engineer licensed under this ordinance shall within the first 10 da. of January and July, respectively, of each year, make a written report to the board of examiners of the condition of the engine, boilers, and steam apparatus under his charge, and every boiler or water tender shall at the same time make a similar report of the condition of the boiler or boilers under his charge.

Engineers in charge of locomotives are exempt from the provisions of this ordinance. as are also all boilers used for heating private dwellings and hothouses and other boilers carrying a pressure of not more than 10 lb. per sq. in., and the persons operating them.

Philadelphia, Pa.—It is unlawful for any person or persons to have charge of or to operate a steam boiler or steam engine over 10 H. P. in cities of the first class of the commonwealth of Pennsylvania, except locomotive boilers used in transportation and steam engines and steam boilers carrying a pressure of less than 15 lb. per sq. in., unless said person or persons are upward of 21 yr. of age and hold a license as provided for by the laws.

The laws state that all persons desiring authority to perform the duties of an engineer shall apply to the boiler inspector of such cities, who shall examine the applicant as to his knowledge of steam machinery and his experience in operating the same, also the proofs he produces in support of his claim; and if the inspector is satisfied that the applicant's character, knowledge, and experience in the duties of an engineer are such as to authorize the belief that he is a suitable person to be entrusted with the powers and duties of such station, he shall grant him a license, on the payment of $3, authorizing him to be employed in such duties for the term of 1 yr., and such license shall be annually renewed, without examination, upon the payment of $1, provided it is presented for renewal within 10 da. after its expiration.

Licenses so granted shall be graded into two classes, one of which shall entitle the licensee to have charge of or to operate stationary steam boilers and steam engines only, and the other of which shall entitle the licensee to have charge of or to operate portable steam boilers and steam engines only. Such licenses shall not be transferred from one grade to the other without a reexamination, said reexamination to be conducted without cost to the licensee.

No person shall be eligible for examination for license unless he furnishes proof that he has been employed about a steam boiler or steam engine for a period of not less than 2 yr. prior to the date of application, which must be certified to by at least one employer and two licensed engineers.

The inspector shall have authority to suspend or revoke licenses upon proof of negligence or incompetency of the holders of licenses. Licenses must be framed under glass and placed in a conspicuous place about the engine or boiler rooms. When a licensed engineer vacates a position, he must notify the boiler inspector of that fact.

Pittsburg, Pa.—See Pennsylvania state laws, on page 250.

Rochester, N. Y.—The ordinance governing the granting of engineers' licenses in Rochester, N. Y., is in part as follows: No person shall operate any boiler to generate steam within the city of Rochester, except for railroad locomotive engines and for heating purposes in private dwellings, unless he be 21 yr. of age and shall have been duly examined and licensed for that purpose as required by the terms of this ordinance.

The common council of the city shall appoint a committee of three competent persons to examine all applicants for license to operate steam boilers and to issue licenses to such applicants as shall be found qualified.

Applications for examinations shall be made in writing to the city clerk, and must state the location and capacity of the boiler plant that the applicant intends to operate. Every application must be accompanied by a certificate of two reputable persons to the effect that the applicant is of good character.

The examining committee shall hold meetings at least twice each month for the transaction of business and the conduct of examinations of engineers. In case an applicant upon his first examination shall fail to satisfy the committee of his ability to operate the boiler plant mentioned in his application, a temporary permit may be granted to him to operate the plant for a period not exceeding 20 da., at which time the applicant must again present himself for examination. Such temporary permit shall not be granted more than once to the same person.

Every person found qualified by the committee to operate a steam boiler shall be entitled to receive a license for that purpose. The fee for such license is $2, to be paid to the city treasurer. Licenses expire at the end of the year from the date of issuance, and may be renewed for a term of 1 yr. on the payment to the city treasurer of $1. In case of a change of position, licensed engineers, under this act, shall notify the city clerk within 1 wk. of such change and present themselves for examination for the new position. If found qualified for the new position, a license to operate the new plant for the unexpired part of the year covered by the original license shall be issued without further fee.

The common council of the city of Rochester has the power to suspend or revoke a license at any time upon proof of negligence or incompetency.

Santa Barbara, Cal.—A board of examining engineers, consisting of three members appointed by the mayor, administers and enforces the license laws in the city of Santa Barbara, Cal. It is the duty of the board to examine all applicants for engineer's license. The board shall hold one meeting a month— on the second Tuesday—for the purpose of examining applicants for license.

Licenses are granted for a term of 1 yr. upon a payment of $3 from each applicant. They may be revoked by the board upon proof of incompetency or neglect on the part of the licensee. Where boilers are used and engines are run night and day, the owner or the user must employ at least two licensed engineers, who may stand watch alternately. Boilers or steam-generating apparatus of 5 H. P. or over must be operated by a competent engineer who has secured a license. Automobiles are exempt from the provisions of this ordinance, as are also all engines and boilers of locomotives used on railroads.

Every applicant for a license who fails to pass the examination of the board shall be required to wait 4 wk. before making another application. Any applicant who fails to pass after the third trial shall not be permitted to appear again before the board for 4 mo.

Engineers shall notify the boiler inspector when they enter any employment as an engineer, and within 3 da. after give the name of the employer and the location of the boiler or other apparatus in charge. They also shall report the condition of the apparatus under their charge semiannually, during the first 3 da. of January and July of each year. Licenses must be framed under glass and placed in a conspicuous place near the boilers or engines.

St. Joseph, Mo.—A board of engineers consisting of the city boiler inspector and two mechanical engineers, all of whom are appointed by the mayor, with the consent of the common council, administers and enforces the license laws in St. Joseph, Mo.

The board convenes for business twice each month to examine into the qualifications of applicants for engineer's license. The law states that the board shall grant licenses, charging each applicant the sum of $10 for engineer's license, $3 to be deposited with the clerk of the board when application is made. Each applicant is to be allowed three trials. If he then fails to pass a satisfactory examination, the applicant shall then forfeit the money deposited with the clerk of the board; but if the applicant successfully passes the examination, the board shall grant him a license for the term of 1 yr. upon the payment of an additional $7. The fee for each annual renewal of license shall be $2.50.

Any person taking charge of a steam boiler or steam boilers, for heating purposes only, shall be examined by the board of engineers, and if found qualified the board shall grant him a license to that effect on the payment of $5, said license to be in force for 1 yr. from the date of issue. Such licenses may be reissued upon the payment of a fee of $2. The board has the power to revoke licenses upon proof of incompetency or neglect.

When boilers are used and engines run night and day, two licensed engineers must be employed and must stand watch alternately.

All engineers, engines, and boilers of the fire department of St. Joseph and all steam rollers, steam automobiles, and portable boilers used on the streets of the city are subject to the provisions of this ordinance, except that there shall be no fee charged for the inspection of boilers of fire engines; but all locomotive boilers used on railroads and steam boilers supplied with water automatically, when used for heating dwellings and not carrying a pressure of over 10 lb. per sq. in., are exempt from the provisions of this ordinance.

Every applicant for license who fails to pass the examination is required to wait 2 wk. before again making application for license. If applicants fail to pass after the third trial, they may not make application again for 6 mo.

Engineers must notify the boiler inspector upon accepting employment as an engineer; and, within 3 da. after, he must give the name of his employer and the location of the boiler or boilers under his care. Renewals for license shall be made not later than the first regular meeting of the board next following the expiration of the license. Unless this provision is complied with, it shall be necessary for the applicant to be reexamined, to take out a new license, and to pay the regular fee before referred to.

St. Louis, Mo.—In the city of St. Louis the engineers' license laws are administered and enforced by a board of engineers consisting of the city inspector of boilers and elevators and two other members, all of whom are appointed by the mayor, with the consent of the council. This board convenes for business once in each week to examine applicants for engineers' license.

According to the laws, the board shall grant certificates of license for 1 yr. from the date thereof to all applicants who pass the required examination and satisfy the board as to their fitness. Each applicant for license shall at the time of filing his application, pay to the inspector of boilers and elevators a fee of $2, for each examination, but no charge is made for renewals. Licenses may be revoked by the board, upon proof of incompetency or neglect on the part of the licensee.

The owners or users of steam boilers or engines of a capacity of not over 75 sq. ft. of heating surface, and a steam pressure of not over 25 lb. per sq. in., used for power only, and all boilers under a pressure of 15 lb. per sq. in. used for heating purposes only, shall apply for a permit to employ a competent, careful, and trustworthy person, instead of a licensed engineer; such person to be recommended by two citizens, one of whom shall be a steam user or a licensed engineer, and if found competent by the inspector of boilers and elevators, said permit shall be granted. The inspector of boilers and elevators may revoke such permit for cause.

At all times when boilers are in use and engines run there shall be in charge an engineer having a certificate of license from the board of engineers, such certificate to be displayed in some prominent place where the boilers or engines are in use.

The engineers, engines, and boilers of the fire department, locomotive boilers used on railroads, and steam boilers supplied with water automatically, and having no pumps or injectors and used only for heating dwellings, not carrying a steam pressure of more than 8 lb. per sq. in., are exempt from the provisions of this ordinance.

Every applicant for license who fails to pass the examination of the board is required to wait 4 wk. before again making application for license. Applications for license must be made upon blanks furnished by the inspector of boilers and elevators. The applicant must have had an experience of at least 2 yr. at mechanical or steam engineering, and must write and state his experience on said blank. He shall go before the inspector of boilers and elevators and make oath that his statements are true.

Licensed engineers must give to the inspector of boilers and elevators notice of changes of employment when he accepts or leaves his position, and within 24 hr. thereafter he must give the name of his employer and the location of the boilers in his charge. Licensed engineers must make semiannual reports of the condition of all apparatus under their charge to the inspector of boilers and elevators during the first 10 da. of January and July.

Scranton, Pa.—See Pennsylvania state laws, on page 250.

Sioux City, Ia.—A board of examining engineers consisting of three persons, all of whom are appointed by the mayor, with the consent of the city council, administers and enforces the license laws in Sioux City, Ia.

It is the duty of the board of examining engineers to grant licenses of the first, second, and third grade to persons examined and found qualified. A first-class license qualifies the holder to take charge of boilers of 125 H. P. or over; a second-class license, to take charge of boilers of 25 H. P. or over, up to 125 H. P.; and a third-class license, to take charge of all boilers of less than 25 H. P., 12 sq. ft. of heating surface to constitute 1 H. P. The certificates so granted shall run for 1 yr., at which time they shall be renewed. The board of examiners has the power to revoke licenses upon proof of incompetency or negligence on the part of licensees.

Steam boilers or engines of a capacity of not over 75 sq. ft. of heating surface with a pressure of not over 25 lb. per sq. in., and all boilers not exceeding 75 sq. ft. of heating surface under a pressure of 15 lb. per sq. in. used for heating only, require a person with a permit only, instead of an engineer holding a license. Such person must be recommended by two citizens of Sioux City, one of whom shall be an engineer holding a license from the board of engineers. Permits are granted by the board to the proper persons.

When any boiler or engine requiring the services of an engineer holding a first-class license is run day and night, the owner or user may employ an engineer holding a second-class license, not exceeding 12 hr. at a time, under the instructions of an engineer in charge, holding a first-class license.

Applications for license must be made on blanks furnished by the board of examining engineers. Licensed engineers must notify the board when they accept or leave employment, and within 10 da. after, the name of employer and the location of the boilers must be given. Applications for renewals of license or permits shall be made not later than the third week preceding their expiration.

In case of the failure of any applicant for a license to pass the examination, he may within 10 da. after receiving notice

20

of such failure make written application to the mayor of the city and also to the board of examining engineers for a second examination, which shall be granted by the board within 10 da. after such application is made. Failure to comply with the foregoing requirements as to a second examination necessitates the applicant waiting the pleasure of the board as to when he may be again examined, such time, however, must not exceed 6 mo.

No person shall be granted a first-class license until he gives satisfactory proof that he has had an experience of 5 yr. in steam engineering. No person shall be granted a second-class license until he gives satisfactory proof that he has had an experience of 3 yr. in steam engineering. No person shall be granted a third-class license until he gives satisfactory proof that he has had an experience of 2 yr, in steam engineering. For all the grades of licenses, the sum of $3 each shall be charged and for each annual renewal of licenses the sum of $1 shall be charged by the board of examining engineers. The provisions of this ordinance shall not apply, and shall not be construed to be applicable to, boilers or steam generators of any kind used for heating private residences only, nor to persons in charge of them.

Spokane, Wash.—The board of public works administers and enforces the license laws in the city of Spokane.

Every engineer must appear before the examiner appointed by the board of public works for examination as to qualifications to act as engineer in the city of Spokane. Applicants for examination as engineer shall pay the examiner a fee of 50c. Every person receiving a permit from the board of public works to run or operate a steam engine or boiler, before acting as such, shall pay to the city treasurer a semiannual license fee of $1, and no license shall be granted for a shorter period. Upon presenting the treasurer's receipt for the fee and the permit from the board of public works to the comptroller, that officer shall issue a license for a period of 6 mo.

All engineers shall instruct all night watchmen, or other persons whose duty it may be to get up steam, as to their duty and the practical mode of procedure, but no night watchmen or other person not a licensed engineer shall be permitted by reason of this clause to operate continuously any steam engine or boiler to which this ordinance applies. Applications for license must be made in writing to the board of public works. Nothing in this ordinance shall apply to locomotive engines or boilers.

Tacoma, Wash.—In the city of Tacoma a board of examiners consisting of the city boiler inspector and two other persons all of whom are appointed by the mayor, administers and enforces the engineers' license laws.

Any person desiring to procure a license as a stationary engineer may apply for it to the boiler inspector upon an application blank furnished by the inspector. He shall then be examined as to his qualifications as a stationary engineer by the examiners. At the conclusion of the examination, the board of examiners shall transmit to the city council all papers used in the examination, including all questions and answers given, together with the recommendation of the board as to its findings. The city council decides as to whether or not a license shall be granted the applicant.

Licenses are for a period of 1 yr. and the fee for a license is $2, payable to the city treasurer. Upon the expiration of a license and upon the payment of a fee of $1 to the city treasurer, the board of examiners may recommend a renewal of the same, without further examination of the applicant, for a period of 1 yr.

The licenses granted are classified as follows: First-class or chief engineer's license which entitles the holder to take charge of and control the operation of any steam plant in the city of Tacoma; second-class, or assistant engineer's license, which entitles the holder to take charge of and control the operation of any steam plant in the city not exceeding 150 boiler H. P., or to act as assistant engineer to the chief engineer of any steam plant in the city; third-class engineer's license, which entitles the holder to take charge of and control the operation of any steam plant in the city not exceeding 50 boiler H. P., or to act as an assistant engineer to an engineer of the second class or as a second assistant engineer to an engineer of the first class; fourth-class, or special-engineer's license, which entitles the holder to take entire charge and control of a particular steam plant for which the same license is granted, as stated upon its face.

This ordinance does not apply to locomotive or marine engineers, nor to persons having charge of steam boilers or apparatus used in private dwellings for heating purposes only, in which the steam pressure does not exceed 10 lb. per sq. in. while in operation.

Terre Haute, Ind.—A board of examining engineers consisting of three members appointed by the mayor, subject to confirmation by the common council, administers and enforces the license laws in the city of Terre Haute.

Any person desiring to operate or to have charge of any steam boiler or steam-generating apparatus under steam pressure within the meaning of this act shall make a verified application to the board of examiners for a license to do so. The application shall be in writing, giving particulars as to experience, and attached thereto shall be a certificate of at least two freeholders or householders to the effect that the applicant for license is known to be a fit person to have charge of steam-generating apparatus. The applicant must pay $3 to the board of examiners before the examination is entered upon. If the applicant is found qualified, a license is issued to him.

Applicants must be American citizens at least 21 yr. of age and must have had at least 2 yr. experience in firing stationary steam boilers or steam-generating apparatus, or must have an actual experience of at least 1 yr. in the operation, management, and control of steam engines or boilers or steam-generating apparatus.

Examinations shall be made in writing, except by agreement between the applicant and the board. All grading of such examinations shall be upon the percentage basis. Three grades of licenses—namely, first-class, second-class, and third-class—are issued. Third-class licenses authorize the holder only to operate a steam engine, engines, or other steam-generating apparatus under the direct supervision of a person or persons holding a first-class or a second-class license.

An applicant who has had 5 yr. or more of actual experience as an engineer, and who is otherwise qualified, and who, on examination, makes a grade of 90% or more, shall be entitled to a first-class license. An applicant who is otherwise qualified, and who makes 75% or more, and less than 90%, shall be entitled to a second-class license. An applicant who is otherwise qualified, and who makes 60% or more, and less than 75%, shall be entitled to a third-class license. The board has the power to suspend or revoke licenses upon proof of incompetency or neglect.

All licenses must be renewed 1 yr. from date of issue, and the fee for such renewal is $2. If any licenses are not renewed within 10 da. after the expiration of such, the fee for renewal shall be $3.

Engineers in charge of locomotives are exempt from the provisions of this ordinance, as are also persons who operate boilers used for heating purposes only, not carrying more than 15 lb. of steam pressure per square inch.

Washington and District of Columbia.—In the District of Columbia, which includes the city of Washington, all persons applying for a steam engineer's license shall be examined by a board of examiners composed of the boiler inspector of the District of Columbia and two practical engineers, to be appointed by the district commissioner.

Applicants for license must be at least 21 yr. of age and of good character, and certified to by at least three citizens of the District of Columbia, themselves of good character. Application must be made in writing. The fee for a license as steam engineer is $3. Licenses may be revoked upon proof of negligence on the part of licensee.

It is unlawful for any person to act as a steam engineer who is not regularly licensed to do so by the commissioners of the District of Columbia. Boilers, and operators of same used for heating purposes, where the water returns to the boiler without the use of a pump, injector, or inspirator, are exempt from the provisions of this act. Engineers licensed by the United States government are also exempt.

Yonkers, N. Y.—In the city of Yonkers, N. Y., a board of examiners consisting of three members appointed by the mayor, with the consent of the common council, administers and enforces the license laws.

According to the law, it shall be the duty of the board of examiners to examine all persons proposing to operate, manage, or run steam boilers or engines, and to certify the qualifications of such as are found competent as either first- or second-class engineers. Persons desiring examination shall apply to that member of the board officially known as the inspector of engineers and of steam boilers in the city of Yonkers. Application must be in writing, specifying in full the experience of the applicant and stating where, how and by whom employed; also, the application must contain the names of at least three residents of the city who can vouch for the good character of the applicant. Upon receiving such application, the inspector notifies the board of examiners, who in turn notifies the applicant as to the time and place of examination.

There shall be two classes of licenses issued—one to first-class engineers and one to second-class engineers.

A license for a second-class engineer shall authorize the person to whom it is issued to have charge of a steam boiler only when not connected with any engine, or when the engine connected therewith shall not be running or in operation. All licenses shall be for 1 yr. from the date of issue. Licenses may be renewed by the board for a period of 1 yr.

No license will be issued by the board to any person who lacks ability, knowledge or experience, and who has not a good character.

Licenses may be revoked by the board upon proof of neglect or incompetency.

For the examination of engineers, the sum of $2 must be paid by each applicant, and for each renewal of same, 50c.

This ordinance does not apply to railroad locomotives used as such, nor to boilers actually used for propelling steam vessels navigating the Hudson River, when the same shall have been inspected and licensed according to the United States laws; nor does it apply to steam boilers used solely for heating purposes, or the persons operating them.

EXTRACTS FROM UNITED STATES LAWS

Following is a true copy of certain sections from the United States Marine Laws, the sections relating to the licensing of marine engineers:

"1· Before an original license is issued to any person to act as a master, pilot, or engineer, he must personally appear before some local board or a supervising inspector for examination; but upon the renewal of such license, when the distance from any local board or supervising inspector is such as to put the person holding the same to great inconvenience and expense to appear in person, he may, upon taking oath of office before any person authorized to administer oaths, and forwarding the same, together with the license to be renewed, to the local board or supervising inspector of the district in which he resides or is employed, have the same renewed by the said inspectors, if no valid reason to the contrary be known to them; and they shall attach such oath to the stub end of the license which is to be retained on file in their office: Provided, however, That any officer holding a license, and who is engaged in a service which necessitates his continuous absence from the United States, may make application in writing for one renewal and transmit the same to the board of local inspectors with a statement of the applicant, verified before a consul or other officer of the United States authorized to administer an oath, setting forth the reasons for not appearing in person, and upon receiving the same the board of local inspectors that originally issued such license shall renew the same for one additional term of such license, and shall notify the applicant of such renewal.

" The first license issued to any person by a United States inspector shall be considered an original license, where the United States records shows no previous issue to such applicant."

" No original license shall be issued to any naturalized citizen on less experience in any grade than would have been required of an American by birth.

" 2. All licenses hereafter issued to masters, mates, pilots, and engineers shall be filled out on the face with pen and ink instead of typewritten. Inspectors are directed, when licenses are completed, to draw a broad pen and black ink mark through all unused spaces in the body thereof, so as to prevent as far as possible, illegal interpolation after issue.

"3· Licensed officers serving under 5 yr. license, entitled by license and service to raise of grade, shall have issued to them new licenses for the grade for which they are qualified, the local inspectors to forward to the Supervising Inspector-General the old license when surrendered, with the report of the circumstances of the case.

"But the grade of no license shall be raised, except as hereinafter provided, unless the applicant can show 1 yr. actual experience in the capacity for which he has been licensed.

"4· In case of loss of license, of any class, from any cause, the inspectors, upon receiving satisfactory evidence of such loss, shall issue a certificate to the owner thereof, which shall have the authority of the lost license for the unexpired term, unless in the meantime the holder thereof shall have the grade of his license raised after due examination; in which case a license in due form for such grade may be issued.

"5· Inspectors shall, before granting an original license to any person to act as an officer of a vessel, require the applicant to make his written application upon the blank form authorized by the Board of Supervising Inspectors, which application shall be filed in the records of the Inspectors' office. Inspectors shall also, when practicable, require applicants for pilot's license to have the written indorsement of the master and engineer of a vessel upon which he has served, and of one licensed pilot, as to his qualifications. In case of applicants for original engineer's license, they shall also, when practicable, have the indorsement of the master and engineer of a vessel on which they have served, together with one other licensed engineer.

"6· No original master's, mate's, pilot's, or engineer's license shall be issued hereafter or grade increased except upon written examination, which written examination shall be placed on file as records of the office of the inspectors issuing said license; and, before granting or renewing a license, inspectors shall satisfy themselves that the applicants can properly hear the bell and whistle signals.

"7· Any applicant for license who has been duly examined and refused may come before any local board for reexamination after 1 yr. has expired.

"8· When any person makes application for license it shall be the duty of the local inspectors to give the applicant the required examination as soon as practicable.

"9· Any person who has served at least 1 yr. as master, commander, pilot, or engineer of any steam vessel of the United States in any service in which a license as master, mate, pilot, or engineer was not required at the time of such service, shall be entitled to license as master, mate, pilot, or engineer, if the inspectors, upon written examination, as required for applicants for original license, may find him qualified: Provided, That the experience of any such applicant within 3 yr. of making application has been such as to qualify him to serve in the capacity for which he makes application to be licensed.

"Any officer of the Naval Militia who is an applicant for license as chief engineer or assistant engineer of steam vessels of the Naval Militia may be examined by inspectors and granted a special license as such, and for no other purpose, if, in the judgment of the inspectors, he is qualified. And the inspectors shall state on the license the name of the vessel on which such master, mate, pilot, or engineer is authorized to act in the capacity for which he is licensed.

"All licenses issued to officers of the Naval Militia provided for in the preceding paragraph of this section shall be surrendered upon the party holding it becoming disconnected from the Naval Militia by resignation or dismissal from such service; and no license shall be issued as above except upon the official recommendation of the chief officer in command of the Naval Militia station of the State in which the applicant is serving.

"10· No person holding special license (Form 878) shall be eligible for examination for a higher grade of license until such person has actually served two full seasons under the authority of his license and one additional full season in a subordinate capacity upon steamers requiring regularly licensed officers.

"11· Whenever an officer shall apply for a renewal of his license for the same grade the presentation of the old certificate shall be considered sufficient evidence of his title to renewal, which certificate shall be retained by the inspectors upon their official files as the evidence upon which the license was renewed: Provided, That it is presented within 12 mo. after the date of its expiration, unless such title has been forfeited or facts shall have come to the knowledge of the inspectors which would render a renewal improper; nor shall any license be renewed in advance of the date of the expiration thereof, unless there are extraordinary circumstances that shall justify a renewal beforehand, in which case the reasons therefor must appear in detail upon the records of the inspectors renewing the license.

"12· When the license of any master, mate, pilot, or engineer is revoked such license expires with such revocation, and any license subsequently granted to such person shall be considered in the light of an original license. And upon the revocation or suspension of the license of any such officer said license shall be surrendered to the local inspectors ordering such suspension or revocation.

"13· The suspension or revocation of a joint license shall debar the person holding the same from the exercise of any of the privileges therein granted, so long as such suspension or revocation shall remain in force.

"14· When the license of any master, mate, pilot, or engineer is suspended, the inspectors making such suspension shall determine the term of its duration, except that such suspension cannot extend beyond the time for which the license was issued.

CLASSIFICATION OF ENGINEERS

"20· Chief engineer of ocean steamers.

"Chief engineer of condensing lake, bay, and sound steamers.

"Chief engineer of non-condensing lake, bay, and sound steamers.

"Chief engineer of condensing river steamers.

"Chief engineer of non-condensing river steamers.

"Any person holding chief engineer's license shall be permitted to act as first assistant on any steamer of double the tonnage of same class named in said chief's license.

"Engineers of all classifications may be allowed to pursue their profession upon all waters of the United States in the class for which they are licensed.

First Assistant

"First assistant engineer of ocean steamers.

"First assistant engineer of condensing lake, bay, and sound steamers.

"First assistant engineer of non-condensing lake, bay, and sound steamers.

"First assistant engineer of condensing river steamers.

"First assistant engineer of non-condensing river steamers.

"Engineers of lake, bay, and sound steamers, who have actually performed the duties of engineer for a period of 3 yr., shall be entitled to examination for engineer of ocean steamers, applicant to be examined in the use of salt water, method employed in regulating the density of the water in boilers, the application of the hydrometer in determining the density of sea water, and the principle of constructing the instrument; and shall be granted such grade as the inspectors having jurisdiction on the Great Lakes and seaboard may find him competent to fill.

"Any assistant engineer of ocean steamers of 1,500 gr. T. and over, having had actual service in that position for 1 yr. may, if the local inspectors, in their judgment, deem it advisable, have his license indorsed to act as chief engineer on lake, bay, sound or river steamers of 750 gr. T. or under.

"Any person having had a first assistant engineer's license for 2 yr. and having had 2 yr. experience as second assistant engineer, shall be eligible for examination for chief engineer's license.

Second Assistant

"Second assistant engineer of ocean steamers.

"Second assistant engineer of condensing lake, bay, and sound steamers.

"Second assistant engineer of non-condensing lake, bay, and sound steamers.

"Second assistant engineer of condensing river steamers.

"Any person having had a second assistant engineer's license for 2 yr., and having had 2 yr. experience as third assistant engineer, shall be eligible for examination for first assistant engineer's license.

Third Assistant

"Third assistant engineer of ocean steamers.

" Third assistant engineer of condensing lake, bay, and sound steamers.

"First, second, and third assistant engineers may act as such on any steamer of the grade of which they hold license, or as such assistant engineer on any steamer of a lower grade than those to which they hold a license.

"Any person having a third assistant engineer's license for 2 yr., and having had 2 yr. experience as oiler or water tender since receiving said license, shall be eligible for examination for second assistant engineer's license.

"Inspectors may designate upon the certificate of any chief or assistant engineer the tonnage of the vessel on which he may act.

"Any assistant engineer may act as engineer in charge on steamers of 100 T. and under. In all cases where an assistant engineer is permitted to act as engineer in charge, the inspectors shall so state on the face of his certificate of license without further examination.

"21· It shall be the duty of an engineer when he assumes charge of the boilers and machinery of a steamer to forthwith thoroughly examine the same, and if he finds any part thereof in bad condition, caused by neglect or inattention on the part of his predecessor, he shall immediately report the facts to the master, owner, or agent, and to the local inspectors of the district, who shall thereupon investigate the matter, and if the former engineer has been culpably derelict of his duty, they shall suspend or revoke his license.

"22· Before making general repairs to a boiler of a steam vessel the engineer in charge of such steamer shall report, in writing, the nature of such repairs to the local inspector of the district wherein such repairs are to be made.

"And it shall be the duty of all engineers when an accident occurs to the boilers or machinery in their charge tending to render the further use of such boilers or machinery unsafe until repairs are made, or when, by reason of ordinary wear, such boilers or machinery have become unsafe, to report the same to the local inspectors immediately upon the arrival of the vessel at the first port reached subsequent to the accident, or after the discovery of such unsafe condition by said engineer.

"23· Whenever a steamer meets with an accident involving loss of life or damage to property it shall be the duty of the licensed officers of any such steamer to report the same in writing and in person without delay to the nearest hoard. Provided, That when from distance it may be inconvenient to report in person it may be done in writing only and the report sworn to before any person authorized to administer oaths.

"24· No person shall receive an original license as engineer or assistant engineer (except for special license on small pleasure steamers and ferryboats of 10 T. and under, sawmill boats, pile drivers, boats exclusively engaged as fishing boats, and other similar small vessels), who has not served at least 3 yr. in the engineer's department of a steam vessel, a portion of which experience must have been obtained within the 3 yr. next preceding the application.

"Provided, That any person who has served 3 yr. as apprentice to the machinist trade in a marine, stationary, or locomotive engine works, and any person who has served for a period of not less than 3 yr. as a locomotive or stationary engineer, and any person graduated as a mechanical engineer from a duly recognized school of technology, may be licensed to serve as an engineer of steam vessels after having had not less than 1 yr. experience in the engine department of steam vessels, a portion of which experience must have been obtained within the 3 yr. preceding his application; which fact must be verified by the certificate in writing, of the licensed engineer or master under whom the applicant has served, said certificate to be filed with the application of the candidate; and no person shall receive license as above, except for special license, who is not able to determine the weight necessary to be placed on the lever of a safety valve (the diameter of valve, length of lever, distance from center of valve to fulcrum, weight of lever, and weight of valve and stem being known), to withstand any given pressure of steam in a boiler, or who is not able to figure and determine the strain brought on the braces of a boiler with a given pressure of steam, the position and distance apart of braces being known, such knowledge to be determined by an examination in writing, and the report of examination filed with the application in the office of the local inspectors, and no engineer or assistant engineer holding a license shall have the grade of the same raised without possessing the above qualifications. No original license shall be granted any engineer or assistant engineer who cannot read and write and does not understand the plain rules of arithmetic.

"25. Any person may be licensed as engineer (on Form 2130⅛) (New Form 880) on vessels propelled by gas, fluid, naphtha, or electric motors, of 15 gr. T. or over, engaged in commerce, if in the judgment of the inspectors, after due examination in writing, he be found duly qualified to take charge of the machinery of vessels so propelled.

"Any person holding a license as engineer of steam vessels, desiring to act as engineer of motor vessels, must appear before a board of local inspectors for examination as to his knowledge of the machinery of such motor vessels, and if found qualified shall be licensed as engineer of motor vessels. Form 878, special license to engineers, shall be issued only to engineers in charge of vessels of 10 T. and under. All other licenses to engineers shall be issued on Forms 876 and 877, according to grades specified in this section."

MEMORANDA

MEMORANDA

Promotion
Advancement in Salary
and
Business Success

Secured
Through the

STEAM ENGINEERING

Steam-Electric
Engine Running
Stationary Firemen's
Engine and Dynamo Running
Locomotive Running

COURSES OF INSTRUCTION

OF THE

International
Correspondence Schools

International Textbook
Company, Proprietors

SCRANTON, PA., U. S. A.

SEE FOLLOWING PAGES

Salary Increased 450 Per Cent.

For several years I had charge of small steam plants which paid me about $20 a month. An I. C. S. Representative put the correspondence school question up to me in such a way I felt it plainly marked "Opportunity." I subscribed for the Engine Running Course, and began studying in my spare moments. Night after night, when the little ones were in dreamland, I was digging into the technicalities of steam engineering.

My reward is being gathered each week as I receive my pay envelope, increased by 450 per cent. over what it was when I bought the Course.

I now have charge of a new steam-electric plant of about 850 horsepower, having from five to eight men in my department. The position I fill was won only by the preparation the I. C. S. gave me. I know of no other way I could have prepared myself for this position. I have my diploma from the I. C. S., and I prize it more highly than any mark of honor I have received.

To all ambitious men I say, **"To succeed you need only pluck, push, perseverance, and a Course in the I. C. S."**

J. EDGAR WILLIAMS,
315 Cable St., Indianapolis, Ind.

IGNORANCE *HINDERS*; THE I. C. S. *HELPS*

S. T. RICHARDSON, Grosse Ile, Mich., found himself stuck fast in a night engineer's position, unable to advance because of a defective education. Knowing nothing of fractions, nor even of division, he took up our Complete Steam Engineering Course. By so doing he fitted himself for the position which he now holds as chief engineer at a salary of $125 a month.

150 PER CENT. INCREASE

That it pays an engineer to acquire a first-class knowledge of everything connected with his work is proved by the case of C. W. FELLOWS, First National Bank Building, Houston, Tex. He was earning $70 a month when he enrolled for the Stationary Engineers' Course. He is now general superintendent of a large office building, having 20 men at work under his direction. His salary has increased 150 per cent.

CHOSEN AHEAD OF MANY OTHERS

ELVIN THOMPSON, Mount Vernon, Ohio, had only a little education and was working as a fireman when he enrolled for the Stationary Engineers' Course. After graduating from the Course, he made application for the position of engineer in the State Sanitarium. Although there were hundreds of applicants, he was chosen. He is now considered an authority on steam engineering and his salary is 200 per cent. larger than when he enrolled.

ONE *THOUSAND* DOLLARS A YEAR LARGER

W. A. BERGER, Hotel Rome, Omaha, Neb., was working in a factory when he enrolled for the Stationary Engineers' Course. He praises the I.C.S. because they have advanced him to the position of chief engineer of the Hotel Rome and Hotel Millard, increasing his salary about one thousand dollars a year.

EARNS $150 A MONTH

CLARENCE GRETTUM, Innisfail, Alberta, Can., had obtained so little education that he could barely read and write when he enrolled for the Complete Steam Engineers' Course. At the time he was earning $60 a month working twelve hours a day as a fireman. He has now taken charge of the Municipal Electric Light Plant of his city at a salary of $150 a month.

500 PER CENT. LARGER

When ARNOLD W. RIDLEY, 3256 Madison St., Denver, Colo., enrolled with the I.C.S. for the Steam Engineering Course he was employed as a helper. He declares that a fireman's practical experience combined with the *I.C.S.* training will make an efficient engineer. He is now plant superintendent for the Denver Gas & Electric Co. His salary has increased about 500 per cent.

3

300 Per Cent. Increase

Because of the death of his father, leaving his mother on her own resources, R. J. BISSETT, 814-16 Columbia Building, Cleveland, Ohio, was obliged to leave school and compelled to go to work at the age of 13. While he was earning $2.75 a day as an engineer at the age of 34, he enrolled with the I. C. S. for the Stationary Engineers' Course. One year later he was made superintendent of the Chamberlain & Target Company, and his salary increased. This position he held for about 10 years. For the past 8 years he has been the president of the R. J. Bissett Company, handling steam appliances, having now 20 agents under his direct supervision. His income has increased 300 per cent., and sometimes rises even higher.

PRAISES THE SCHOOLS

C. S. COUPLAND, Box 47, Maricopa, Calif., says that he cannot praise the Schools too highly, since his Steam-Electric Course has advanced him from a position as steamfitter's apprentice to the position of chief engineer, increasing his salary from $50 to $125 a month.

SALARY INCREASED 266 PER CENT.

P. J. GRACE, 602 Atlantic St., Bridgeport, Conn., was a fireman working 12 hours a day in an electric light works when he enrolled with the I.C.S. for the Stationary Engineers' Course. He is now the chief engineer of the plant of the Locomobile Company, having 22 men under his charge. His salary has increased 266 per cent.

EARNS $1,800 A YEAR

C. M. IRWIN, 429 Walnut St., San Francisco, Calif., is now chief engineer of the Head Building. Before he had a technical knowledge of engineering he was a machinist at $3 a day. His I.C.S. Course has raised his salary to $150 a month.

GRADUATE BECOMES CHIEF ENGINEER

ED. BURROW, Box 175, San Angelo, Tex., is proud of his diploma, because it has taken him from a position in the engine room where he worked all night for $24 a month to the place of chief engineer at a salary of $150 a month. He says: "I cannot see how a boy with ambition can keep from enrolling."

145 PER CENT. LARGER

E. E. HUNTER, 1520 N. Broadway, Oklahoma City, Okla., was earning $85 a month at the time of his enrolment for the Steam-Electric Course. He declares that this Course was largely instrumental in securing his advancement to the position of chief engineer at the Oklahoma Gas & Electric Company, with an increase in salary of 145 per cent.

NOW CHIEF ENGINEER

GEO. W. DRENNON, 879 Mead Ave., Oakland, Calif., declares that he has derived much benefit from his Marine Engineers' Course for which he enrolled with the I.C.S. When he began to study he was employed as an assistant engineer. He is now chief engineer on the steamer "Frances" for the A., T. & S. F. R. R. He holds an engineer's license for ocean steamers of any tonnage as the result of spare-time study.

5

800 Per Cent. Increase

At the time of enrolling in the I. C. S. for the Stationary Engineers' Course I had received very little schooling—anything beyond long division was a mystery to me. When I subscribed I sold my watch to make the first two payments. My wages at that time were $8.75 a week. I am now engineer and selling agent for the Westinghouse Company at an increase in salary of 800 per cent. There is not the slightest doubt that the I. C. S. Course was the tidal wave that carried my little boat into a port not dreamed of in the beginning. All men have not the same kind of a boat, but they have some kind of boat, nevertheless. It is up to each individual which chart to use.

JOHN M. NICHOLSON,
165 Broadway, New York, N. Y.

DOUBLED HIS PAY

CLARK DULEY, Allensworth, Tulare County, Calif., was earning $60 a month at the time of his enrolment for the Steam-Electric Course. Although he had never seen a dynamo, at the time he took charge of the electric lighting plant of Bandcn, Ore., he was able to make a record of keeping the lights burning for 3 months with only 10 seconds' trouble. His salary has been increased to $130 a month.

ONCE A LABORER

IRA G. WHIPPLE, 1801 Kentucky St., Lawrence, Kans., was a laborer 27 years old when he enrolled for the Steam-Electric Course. He says that his work with the Schools helped him to get his present position as assistant engineer at the Kansas State University.

NOW CHIEF ENGINEER

JOHN M. MORRISON, Glace Bay, N. S., Canada, was running an air compressor for the Dominion Coal Company when he enrolled with the I.C.S. for the Steam-Electric Course. He recommends our system of teaching which enabled him to become chief engineer for the No. 4 Colliery, one of the leading collieries of the company, with an increase in salary of 25 per cent.

SALARY NEARLY DOUBLED

When H. L. FULLER, Rossford, Ohio, enrolled for the Steam-Electric Course, he was working as general utility man around a plate glass works. Today he is first assistant engineer for the Edward Ford Plate Glass Company, in charge of a modern turbine plant at a salary of $100 a month.

FROM $660 TO $1,500 A YEAR

GARRETT BURGESS, 269 Stanford Ave., Detroit, Mich., was working as a helper, earning $660 a year at the time he enrolled for the Engine Running Course. Since obtaining his diploma he has been advanced to the position of assistant chief engineer for the Morgan & Wright Rubber Company at a salary of $1,500 a year.

300 PER CENT. LARGER

EARNEST LEWIS, 5 Mott Ave., Burlington, N. J., was employed as a fireman by the Thomas Devlin Mfg. Company at the time of his enrolment for the Complete Steam Engineering Course. Sixteen months later, through faithful study of his Course, he was advanced to the position of chief engineer, with an increase in salary of 300 per cent. He declares positively that he could never hold his present position if it had not been for the I.C.S.

7

Salary Quadrupled

When I enrolled with the I. C. S. for the Stationary Engineers' Course, I was a machinist earning $13 a week. When a boy at school I could see no need of an education, and quit as soon as the law allowed, having learned very little while I did attend school. For a few years I was satisfied, and then I began to wish that I had something better, but soon found out that I was not qualified for advancement, as I had wasted my time in school. I then got out the old arithmetic, but could not get any satisfaction out of it, so I threw it up and decided that I was a hopeless case. Some time afterward, one of my shop mates told me that he had taken out the Stationary Engineers' Course with the I. C. S. and invited me to his home to see his books. I then enrol.ed, and, although I have not made any very wonderful strides, I have gained steadily until on May 1, 1910, I was appointed chief engineer of the Springfield Street Railway Power Station at a salary four times what I earned in the iron works. My enrolment with the Schools was the best move I ever made.

W. C. TRACY,
791 Main St., Springfield, Mass.

SALARY *MORE* T*HAN* DOUBLED

R. F. SHANK, 22 Florence Ave., Rosedale, Kans., was working as an oiler for $50 a month when he enrolled for our Steam-Electric Course. He had only obtained a fifth grade common school education, and was obliged to work 12 hours a day. Largely through the study of the Bound Volumes, he advanced himself to the position of chief engineer for the Kimball Cereal Company, of Kansas City, Mo., at a salary of $21 a week.

250 PER CENT. INCREASE

When C. F. RASMUSSEN, Clay Center, Kans., enrolled with the I. C. S. for the Stationary Engineers' Course, he was working in a creamery at $30 a month. Having graduated from our Stationary Engineers' Course, and also from our Steam-Electric Course, he has been able to command a better position, and hence has been advanced from time to time until he now draws 250 per cent. larger salary, as superintendent of the City Light and Water Works plant.

NOW CHIEF ENGINEER

J. G. BLYTHEWOOD, Voth, Tex., was trying to learn the engi neer's profession when he enrolled for our Stationary Engi neers' Course. At that time he was working wherever he could, earning $1 a day. He says that our system is the best way for a laboring man to acquire an education, since it has promoted him to the position of chief engineer for the Beaumont Irrigating Company at a salary of $150 a month.

GRADUATE GAINS 150 PER CENT.

JOHN W. HILFRANK, White Plains, N. Y., is now chief engineer at the New York Orthopædic Dispensary and Hospital. When he enrolled for our Stationary Engineers' Course, he was working for $12 a week. He recommends the Course, from which he has graduated, because it has increased his salary 150 per cent.

NOW FOREMAN

GEORGE KORNEGOE, Van Meter, Iowa, had only a common school education when he enrolled for our Steam-Electric Course. He is now foreman for the Platt Company, brick and tile manufacturers, drawing $100 a month.

NOW EARNS $196 A *MONTH*

ARCHIE F. HUBBARD, Rich Grove, Tulare County, Cal., was running a harvester when he began to study our Engine Running Course. He now runs a 130-horsepower traction engine, earning $196 a month and board during the harvesting months, and $130 a month during the plowing months.

9

Now Superintendent

When I enrolled with the International Correspondence Schools for the Complete Steam Engineering Course I was working on a farm earning $20 a month. I had had a public school education, and not being able to go to the high school, decided to try the I. C. S. method. After completing my Course with very little difficulty, I became interested in the mines. I am now superintendent of the Peterson Lake Mining Company at a salary of about $160 a month. I am now working on your Metal Mining Course. The large increase in my earning capacity is due almost entirely to your splendid Courses, which are worth their cost 10 times over.

HENRY SANKEY,
Cobalt, Ontario, Canada

10

WORKING AGAINST ODDS

CARL SIMPSON, Newark, Ohio, had only a little education and was working 13 hours a day 7 days in a week, trying to support his family when he enrolled for our Engine Running Course. Although his own people thought that he was spending his time and money for nothing, he struggled on alone until he obtained his diploma. He is now chief engineer at the Municipal Water Works for the city of Newark at a salary of $100 a month, nearly double what he received at the time of enrolment.

INCREASED HIS SALARY 100 PER CENT.

C. W. SINGER was earning $12 a week when he subscribed for our Advanced Engine Running Course. He is now superintendent of the Optimo Mining Company's mines at Linden, Wis., having 35 men at work for him, and earning $25 a week.

HAD NO EDUCATION

JOHN G. SCHAFNITZKY, 206 State St., Hudson, N. Y., had never seen the inside of an American school house, and had only been to a German school for 3 years. After picking up a little English from his younger sisters, who were sent to school, he became ambitious to learn and enrolled for our Engine Running Course. At that time he was earning $1.30 for 12 hours' work. He is now master mechanic for the James Stewart Construction Company at a salary of $25 a week straight time.

NOW SUPERINTENDENT

M. C. REYNOLDS, Box 48, Carey, Ohio, was working as a fireman in a small pumping station, earning $35 a month when he telephoned our Representative to come to the works to enroll him for the Engine Running Course. Today he is superintendent of the electric lighting and water works plant of Carey, Ohio, making three times his former salary.

SALARY NEARLY DOUBLED

Forced to leave school at an early age to go to work in the mines, JOHN PARKS, 319 Highland Ave., Lexington, Mo., found himself at the age of 18 with a knowledge of only simple problems in addition. While working as a fireman, he enrolled for the Engine Running Course. He is now engineer at a salary 90 per cent. greater than he received at the time of his enrolment.

HIS COURSE WORTH $50 A MONTH MORE TO HIM

MAYNARD JOSEPH, Collinsville, Ill., was earning $50 a month as a fireman when he took out our Engine Running Course. This enabled him to pass the state examination and to obtain a first-grade certificate of competency as a hoisting engineer. He now has charge of the plant of the Donk Brothers Coal Company, a plant having a capacity of 3,000 tons of coal a day, and his salary has increased $50 a month.

11

Salary Increased 500 Per Cent.

I was employed as a farm hand at $15 a month when I enrolled with the Schools for the Stationary Engineers' Course. This enabled me to become an engineer for the Booth-Kelly Lumber Company, increasing my salary $50 a month. I studied your volumes on electricity, of which I had no previous knowledge, and they enabled me to erect about 75 miles of telephone lines and to install a central office. I am now doing construction work for the Oregon Power Company of this place, at a salary 500 per cent. larger than what I received at the time of enrolment.

PHILIP A. JOHNSON,
Springfield, Ore.

NOW GENERAL MANAGER

JOHN J. PRICE, Cement City, Mich., had only a common school education before he enrolled with the International Correspondence Schools for the Stationary Engineers' Course. He did his studying at night, running an engine during the day. After working for the Peninsular Portland Cement Company in a subordinate position, he was given charge of their power house, and then of their dredges, until he was made general manager of the entire works at a salary of $2,000 a year, the works employing 150 men. Mr. Price earned $50 a month when he enrolled.

NOW CHIEF ENGINEER

D. S. KENNEDY, 19 Wetmore St., Warren, Pa., was earning $65 a month as an electrician when he began to study our Stationary Engineers' Course. He recommends this Course to young men because it has enabled him to become chief engineer of the Warren Electrical Company with an increase of 50 per cent. in salary.

NO LONGER BLUNDERS ALONG

CHARLES H. WAINNER, Box 353, Pratt, Kans., had no regular position, and was working wherever he could at $1.25 a day when he enrolled for our Stationary Engineers' Course. Until he had mastered this, no one wanted to have him blundering around an engine room. He is now a successful engineer in charge of the power plant for the Pratt Milling Company, at a salary 150 per cent. larger than when he first enrolled.

NOW PROPRIETOR

OTIS MORRIS, Warren, Idaho, was making $1.25 a day when he enrolled with the I. C. S. for the Stationary Engineers' Course. He is now in business for himself, leasing several mines, and taking charge of all machinery.

NOW GENERAL MANAGER

JOHN HARRIS, Lilly, Pa., was firing boilers at night when he enrolled for our Engine Running Course. This, he says, proved a great help to him, enabling him to become the general manager of the James Harris & Sons Bituminous Coal Mines.

233 PER CENT. INCREASE

GUS LUNDGREN, Cherokee, Iowa, was a fireman in a small electric lighting plant earning but $30 a month when he enrolled with the I. C. S. for the Stationary Engineers' Course. Without this, he says, he could never have reached his present position as manager of the Cherokee Electric Company. His income has increased 233 per cent.

Now Foreman of Engines

I had been working for the Chesapeake & Ohio Railroad Company as an engineman for several years, and had always been considered an A No. 1 engineer, which made me a little conceited. I imagined that I needed no further training, but since completing my Locomotive Running Course with the I. C. S., I am forced to change my mind, and must now admit that my former knowledge was very limited and ordinary. I would not exchange the benefits I have received from the Course for $1,000 in cash. On October 1, 1911, I was promoted to the position of road foreman of engines, which position I now hold, with headquarters at Covington, Ky.

D. F. EVANS,
1220 Madison St.,
Covington, Ky.

THREE TIMES HIS FORMER SALARY

J. C. WHITTEN, 76 Danforth St., Providence, R. I., while working as a fireman for $12 a week, enrolled for our Locomotive Running Course. Having been a poor boy, he had no chance for an education, needing a dictionary to define common words. His Course enabled him to secure promotion at his first examination. He is now an engineer, earning three times what he received at the time of enrolment.

GRADUATE WORKS FOR THE GOVERNMENT

HUGH L. RUSSELL, Keams Canon, Ariz., was earning $12 a week in a roller mills when he subscribed for our Complete Steam Engineering Course 9 years ago. Since. graduating, he was able to pass the Civil Service examination with a grade of 98.3, receiving immediate appointment. He has since enrolled for the Complete Electrical Engineering Course. He is now a steam and electrical engineer in the service of the government, Interior Department, with a salary which has increased about 300 per cent.

GRADUATE RECEIVES PROMOTION

M. J. McKINNEY, 916 S. Fell Ave., Normal, Ill., finished our Locomotive Running Course and was able to pass the State examination, receiving immediate promotion. He now holds a profitable place as engineer on the Chicago & Alton Railroad.

SALARY MORE THAN DOUBLED

HANS C. BROWN, 14 E. Linden St., Wilkes-Barre, Pa., was a railroad fireman, earning $2.20 a day when he enrolled for our Complete Locomotive Running Course. Although he had received only a common school education in his native land, Denmark, he was able to pass an examination and receive promotion as engineer for the Lehigh Valley Railroad Company, where he now earns more than double his former salary.

SALARY INCREASED $95 A MONTH

When NICHOLAS COLILAR, R. F. D. 31, Costello, Potter County, Pa., enrolled for the Complete Locomotive Running Course, he was earning $45 a month as a fireman. His Course has helped him beyond his expectation. He is now a locomotive engineer, for the Emporium Lumber Company, earning $95 a month more than when he enrolled.

OFTEN EARNS $150 A WEEK

W. R. HAY, Gulfport, Miss., was straw boss of a ditch gang repairing streets, earning $12.25 a week, when he enrolled with the I. C. S. for the Complete Steam Engineering Course. He also studied our Marine Engineering Course and Mechanical Drawing. He is now chief engineer and manager of the tug "Beaver," earning from $90 to $150 a week.

Bettered His Position

When I enrolled with the I. C. S. for the Complete Locomotive Running Course, I was employed as an engine wiper at a salary of $54 a month. Since then I have become fireman, and was later promoted to the position of engineer. I believe that my work with the Schools on the above course helped me a great deal in my examinations enabling me to pass with great success. I am now employed by the Northwestern Pacific Railway Co. as an engineer, averaging from $125 to $160 a month.

G. F. BRADLEY, Jr.,
310 3rd St.,
San Rafael, Cal.

CPSIA information can be obtained at www.ICGtesting.com
Printed in the USA
BVOW06s0127130715

408388BV00010B/226/P